T0295204

IET ENGINEERING SERIES 190

Synchrophasor Technology

Other volumes in this series:

Synchrophasor Technology

Real-time operation of power networks

Edited by
Nand Kishor and Soumya R. Mohanty

The Institution of Engineering and Technology

Published by The Institution of Engineering and Technology, London, United Kingdom

The Institution of Engineering and Technology is registered as a Charity in England & Wales (no. 211014) and Scotland (no. SC038698).

The Institution of Engineering and Technology
Futures Place
Kings Way, Stevenage
Hertfordshire SG1 2UA, United Kingdom

www.theiet.org

British Library Cataloguing in Publication Data
A catalogue record for this product is available from the British Library

ISBN 978-1-83953-284-9 (hardback)
ISBN 978-1-83953-285-6 (PDF)

Typeset in India by MPS Limited

Cover image Own garden/Moment via Getty images

Contents

About the editors

Nand Kishor is a Professor in the Department of Engineering, Østfold University College, Norway. Previous assignments include Marie Curie experienced researcher (Marie Curie fellow) at Aalto University, Finland. He has authored and co-authored over 80 SCI international journal publications and has edited two books titled *Modeling and Dynamic Behaviour of Hydropower Plants* and *ICT for Electric Vehicle Integration with the Smart Grid,* along with Dr Jesus Fraile-Ardanuy. Both books were published by the IET, UK. Presently, he also serves as subject editor/associate editor for *IET Generation, Transmission and Distribution, IET Renewable Power Generation,* and *IEEE Systems Journal.*

Soumya R. Mohanty is an Associate Professor in the Department of Electrical Engineering, Indian Institute of Technology (BHU), Varanasi, India. He is a senior member of the IEEE. His research includes digital signal processing for power system protection, multi-terminal DC, microgrids, wide-area monitoring and control. He has led international research collaborations with Dublin Institute of Technology and Norwegian University of Science and Technology. He has published more than 100 research papers in reputed journals and conferences which includes more than 53 SCI international journals.

Introduction

Phasor measurement units (PMUs) technology provides effective measurement-based approach to support planning and operation of electric grid network. Time-synchronized measurement signals obtained from PMUs can provide better information about grid conditions to the operators. Data-driven approaches for power system monitoring, control and protection can be applied.

In the past one decade, transmission network planners and operators have very well recognized the PMUs for monitoring application. Wide-area visualization has become a reality for transmission network. However, a significant level of improvement is needed to develop approaches towards grid operation, control and protection. Full potential of PMUs data have not been yet established at least for these applications.

On the other hand, distribution network remained simple and necessity for real-time monitoring did not arise. However, in recent years, applying PMUs at distribution network has grown. The distribution-PMUs (D-PMUs) need to have high quality, high speed on distribution-tailored synchrophasor network. In addition, distribution networks generally exhibit more pronounced and erratic variations of operating parameters than transmission networks.

Within this context, the contents of the book include chapters addressed not only in the area of power network monitoring, control but also in the protection.

Chapter 1 presents methods for the detection of oscillation source location and discussion is mainly focussed on low-frequency electromechanical oscillations, natural mode oscillation and forced oscillations. Continuing topics on power oscillations, Chapter 2 discusses on real-time testing in Opal-RT for power oscillation monitoring using algorithm applying mode decomposition on PMUs data.

Chapter 3 presents wide-area control design using predictive approach to damp out low frequency oscillations, sub-synchronous frequency oscillation and voltage oscillation, followed by Chapter 4 presenting control design for low frequency oscillation in reduced-order model of a large power network.

Chapter 5 presents discussion on operation of transmission network, focussing on congestion management in real-time.

Remaining Chapters 6, 7, 8 and 9 in the book present discussion in the area of protection based on synchrophasor data. Chapter 6 presents out-of-step predictive protection approach using future samples. The discussion on fault location in multi-terminal transmission network with analytical derivation of concept is presented in Chapter 7, followed by integrity protection schemes for future power system in Chapter 8. Last Chapter 9 presents generalized application of synchrophasor data for microgrid protection.

Editors
Nand Kishor
Soumya Ranjan Mohanty

Chapter 1

Synchrophasor data for oscillation source location

Bin Wang[1] and Slava Maslennikov[2]

Synchrophasor data have a range of applications, from planning, market operation to reliability operation [1]. Oscillation source location (OSL) is among the most successful synchrophasor applications in today's control rooms, which identifies the equipment destabilizing the system. Having such equipment identified in a timely manner offers real-time, actionable information to grid operators, which is critical for mitigating such dynamic risks and improving system reliability. In the past two decades, many OSL methods have been proposed based on different principles. Synchrophasor data have captured many real-life oscillation events and helped approach a more comprehensive explanation of the oscillation phenomenon in power systems: from natural mode oscillations to forced oscillations, from non-resonance conditions to resonance conditions, from model-based analysis methods to data-driven analysis methods or hybrid methods using both model and data. In addition to the significant progress in both theory and application of OSL, it cannot be denied that all the existing OSL theories are not complete, yet, and there are still rooms to improve the engineering practice of OSL.

This chapter starts with a literature review of existing OSL methods, and then focuses on a promising OSL method, named dissipating energy flow (DEF) method [2], and its latest applications. Specifically, detailed studies on longitudinal power systems [3] are presented to reveal more insights into DEF's practical effectiveness and capability. Subsequently, DEF analysis results on multiple cases from the 2021 IEEE-NASPI Oscillation Source Contest [4] are presented and discussed.

Remark: This chapter mainly focuses on (1) the low-frequency electro-mechanical oscillations, i.e., ranging from 0.1 Hz to ~3 Hz, rather than the high-frequency oscillations involving inverter-based resources, say beyond ~5 Hz, which will only be briefly mentioned; and (2) natural mode oscillations; and (3) forced oscillations that drive the system to oscillate at the forcing frequency and/or its integer multiples, rather than at any non-integer multiples of the forcing frequency due to nonlinearities [5] or bifurcations [6].

[1]Department of Electrical and Computer Engineering, University of Texas at San Antonio, USA
[2]Advanced Technology Solutions, ISO New England, Holyoke, USA

1.1 Review of OSL methods

This section reviews existing OSL methods, classifying them into different categories based on their mechanisms, and discussing their advantages and disadvantages. In the review article [7], many OSL methods proposed by the year 2015 were reviewed and summarized, most of which will not be cited again other than a brief mention. Interested readers are referred to [7] for more details. This section will inherit some main classifications from [7], and focus more on the methods in publications after year 2015. Although the authors have made considerable efforts to be as inclusive as possible when searching for relevant publications, chances are still there that some relevant publications were unintentionally missed. With 70+ publications discussed in this section, it is the authors' hope that the present review could cover all major developments of OSL methods in today's literature.

Similar to [7], this review also only covers the methods identifying the location of oscillation source(s) down to the substation level. Further pinpointing the destabilizing controller and even the destabilizing control block(s) [8–15], though very important as well, are not the focus of this review.

1.1.1 Traveling wave-based methods

It is a fact that any physically realizable systems are causal systems, including power systems. Traveling wave-based OSL methods are based on such a fact and its inference that when a forced oscillation (FO) starts to drive the entire power system to oscillate, it always occurs earlier at locations electrically closer to the source of FO. Especially, the location of the FO source occurs first. In the past, to estimate the arrival time of the oscillation at one location or the difference in arrival time between different locations, the similarity function, and least mean square method were adopted, while the location with the earliest time is identified as the measurement location that is closest to the true oscillation source [7].

A traveling wave-based method relying on the Wiener–Khinchin theorem is used as the reference method in [16]. In this reference method, the cross-correlation of any pair of time series is calculated and used to determine the time difference between the two time series. Then, a consistency check based on triangle inequality is conducted to identify the time series acting as the cause for all others, whose location is identified to infer the oscillation source.

In [17], the FO arrival time is detected by checking whether the Fourier spectrum peaks at the FO frequency. Then, the five phasor measurements units (PMUs) with the smallest FO arrival time are selected to estimate the geographical coordinates of the FO source by a triangulation method. This OSL method is tested with two simulated cases on the 70k-bus Eastern Interconnection system and achieves an error up to 62 miles, while the test on a real case with field data shows an error of 90 miles.

The disturbance arrival time at every available measurement device is estimated in [18] by detecting a sudden change in the singular values of the frequency data matrix with a pre-specified threshold. Along with the geographic coordinates of the measurement devices, locating the disturbance is formulated as a set of algebraic equations with the disturbance's coordinates and the disturbance start time as unknowns. The problem is solved by the temporal scanning algorithm, i.e., solve for disturbance's coordinates with a given disturbance start time. The problem is solved multiple times for variation of the disturbance start time with desirable resolution. Finally, the actual disturbance start time is estimated as the one with a minimum mean squared error by evaluating all measurement samples using the formulated equations. It is worth mentioning that redundant measurements can be easily and naturally included in the formulated equations which may reduce the estimation error. This method is tested with data from 6 frequency disturbance recorders (FDRs) capturing a field event and shows an error of 2.4 degrees in longitude and 0.8 degrees in latitude, which is roughly equivalent to 100 miles. It is also found in this work that the location accuracy is insensitive to errors in the assumed electromechanical wave propagation speed.

Traveling wave-based methods are free of underlying power system model, which only requires the measurements and their location information in the system. In addition, a clean starting-up response of the system is needed, which any other disturbances or noises obfuscating the onset of the FOs are not desired by these methods. A centralized implementation is always needed according to the principle.

1.1.2 Mode shape-based methods

Mode shape-based methods identify the source of oscillation based on the direct comparison of the oscillation magnitude and/or oscillation phase between measurements from different locations. The most commonly seen criterion is to select the location with the largest oscillation magnitude or the leading oscillation phase as the oscillation source. Such a criterion can help correctly locate the oscillation source in many cases. However, it has become well-known that this criterion could fail especially under resonance conditions, i.e., when the FO frequency is equal or close to the frequency of a poorly damped natural mode of the power system. Still, researchers and engineers have been constantly re-visiting the mode shape-based methods, partially because these methods are always data-driven, allowing easy implementations in practice. The following summarizes the recent efforts.

The transfer function between different bus frequencies, i.e., the relative oscillation magnitude and oscillation phase in frequency signals, are used for OSL in [19]. Tests on a 2-area system and a 68-bus system give the following two observations: (1) for high-frequency electromechanical oscillations, a larger oscillation magnitude always means it is closer to the oscillation source; (2) for oscillations around a natural mode frequency, a leading phase means it is closer to the oscillation source.

In [20,21], synchrophasor data are only used to determine the start and end time of detected oscillations. Then, the oscillation source is identified by using SCADA data, which have wide coverage, but has a low resolution and are usually not synchronized. This is not a synchrophasor-based OSL, but is reviewed here due to its relevance. Fortunately, if the oscillations are present in a long-time window and only the oscillation magnitude is evaluated, SCADA data would very likely hit some peaks and valleys of the oscillation and could be used to provide a good estimate of the oscillation magnitude. The work in [20,21] looks at SCADA data measured at all generators, calculates two ranking indices, i.e., the average-crossing counts and the oscillation amplitude measured by the standard deviation of SCADA data, and uses them for OSL.

Mode shape can also be calculated by spectrum analyses. Power spectrum density, excess kurtosis, and cross power spectrum density (CPSD) are employed to find the peak of oscillation at the frequency of interest. Then, the highest peak of CPSD among different machines is selected as the oscillation source, assuming that there is only a single generator-type oscillation source and it is measured or observable [22].

Another two recent papers relying on either the leading oscillation phase [23] or the largest oscillation magnitude (in frequency) [24] for OSL are not detailed here. One interesting phenomenon is observed in [25] that the magnitude of harmonics (of the oscillation frequency) decreases much faster with the increase of distance to the actual oscillation source, as compared to that w.r.t. the base oscillation frequency. Based on this observation, the location with largest magnitude of harmonics is identified to be, or close to, the source of oscillations. As pointed out by [25], such an observation may be distorted if a harmonic frequency resonates with a system's natural oscillation mode, a common limitation possessed by all mode shape-based OSL methods.

1.1.3 Energy-based methods

Energy-based OSL methods are rooted in the Lyapunov function theory and the extension of its computation to network variables based on energy conservation law [26]. Losses in general multi-machine power system models, specifically the resistors, cause difficulties in finding a well-defined energy function [27]. It was reported that resistances in the power networks and loads may either produce or dissipate transient energy, depending on the difference in oscillation phase between voltage and frequency [28]. Paper [29] particularly indicates that a complex inter-action of controls could be a potential reason for the failure of energy-based methods. Mathematical proof has been provided in [30] showing the condition for the potential failure of energy-based methods.

Regardless of the above theoretical difficulties, energy-based OSL methods are being constantly practiced [8,9,31–34] and shown great success [35–37]. Additional improvements for enhancing their capabilities in practical settings include

- Consideration of missing data [38,39].
- Speedup by more robust and accurate modal estimation [39,40].
- Result visualization by the closed contour grouping technique [41].
- Adoption of frequency-domain implementation to avoid challenging filter designs in online environment [42].
- Distributed implementation [43],
- Synchrosqueezed wavelet-based filter for extracting each oscillation mode to be analyzed by the energy-based methods [44].

A few other remarkable improvements are discussed below.

- Almost all existing energy-based methods only rely on the imaginary part of the complex expression of the energy conservation law [26]. Paper [45] directly considers the complex energy function, called CDEF, and proposes to project the CDEF to a direction in the complex domain with the direction defined by $K+j$, where j is the imaginary unit and K is a parameter selected by user. It's observed that a larger K usually leads to an improved performance for identifying the FO source in machine's excitation system, while may possibly worsen the performance in identifying the FO source in machine's governor system.

- A mathematical proof is provided in [46] to show the consistency between the energy and the real part of the eigenvalue corresponding to the observed oscillations. The proof is achieved for arbitrary machine models and for multi-machine power systems. However, how this insight could help better locate the source of forced oscillation has not been revealed. Also, it is interesting that the energy considered in this work contains only one term, corresponding to active power and voltage phase angle, out of the two from the definition of transient energy based on the imaginary part of the energy conservation law [26].

- Real-time network and load models are assumed to be available and incorporated with energy flow calculation in [47]. It would be interesting to see whether/how this could address the difficulties discussed at the beginning of this subsection.

Although the practical effectiveness has been demonstrated, the theory of energy-based OSL methods is still not complete and there are known cases that can cause potential failures of these methods when locating FO sources. The above improvements are either being evaluated for handling the difficult cases, or need to be evaluated. More practical improvements and theoretical investigations are also needed.

1.1.4   Artificial intelligence-based methods

Artificial intelligence (AI) has achieved several good applications in many other fields in the past two decades, but there was only one OSL method reported in [7] based on AI by the year 2015. In recent years, there have been more attempts to apply AI to the OSL problem, which usually differ from each other in the selection of data/feature, neural network/classifier, and training settings among others.

In paper [48], time series of voltage magnitude, voltage angle, rotor angle, and rotor speed are directly and respectively used as input data, while the output is the location of the oscillation source. Support vector machine, decision tree, and random forest are adopted as three classifiers. Different combinations of input data and classifiers are tested. Twenty-one sustained oscillation cases on a 179-bus system are used for training and testing, with additional data generated by varying system loading, fault settings, or measurement errors.

Paper [49] takes bus frequency, active and reactive power injections as the input data, and adopts the deep neural network. Instead of directly working on the time-domain data, this paper first converts the data into the frequency domain, where a Hanning window is adopted to mitigate the leakage effect of the Fourier transform. The test system is also the 179-bus system, where 4553 FO cases are created to form the data set for training and testing, where a high success rate of >98% is achieved.

Standard PMU measurements and their derivatives, including voltage magnitude, voltage angle, bus frequency, active and reactive line flows, are selected as the input data in the work [50]. Multi-variate time series (MTS) matrix is formed for each FO case. The Mahalanobis distance between MTSs (or FO events) is calculated with an assumed coefficient matrix M, which is optimized by minimizing distances between MTSs sharing the same FO source location, and maximizing distances between MTSs associated with different FO source locations. To create a large enough data set for training and testing, load variations, different time delays, and different FO locations are considered to modify a given base case of the test system. A total of 6,000 data samples are created and used for testing the 39-bus system, while 17,400 for the 179-bus system. The success rates of the proposed AI-based OSL method are >97% and >95%, respectively, for the 39-bus and 179-bus systems.

An interesting idea is proposed in [51] which first extracts FO features and converts them into images using the Smooth pseudo Wigner-Ville distribution. Then, the convolutional neural network-based deep transfer learning is applied for OSL. With 2102 (or 500–800) FO samples from the 179-bus system, the accuracy of locating the area (or bus in the area) containing the FO source is 100% (or >93%). With additional 10,000 samples, area (bus) OSL accuracy becomes >99.5% (or >92%).

Another interesting idea is to apply the machine-learning techniques to the results of existing OSL methods, e.g., to the location results by the dissipating energy flow method [52]. To form the training and testing data set, parameters are varied to create different combinations, including the location of FO generator, location of FO load, ambient noise level, oscillation frequency, oscillation magnitude, and load model composition. As a result, 20,000 samples are created on a 240-bus test system [53]. Then, dissipating energy flow method is applied to each of these samples to calculate the dissipating energy flows on all branches, which, along with the true location of FO, are used to train a machine learning-pattern recognition (ML-PR) model. This trained ML-PR will be used as a real-time OSL application. The success rate when testing the 240-bus system is 100%.

1.1.5 Model inference-based methods

The basic idea of a model inference-based method is to use the assumed model (describing the entire system or part of the system such as a single component) and the measured input X to infer the output $\hat{Y}$. The difference between the inferred output $\hat{Y}$ and the measured output Y reflects the accuracy of the assumed model. For power system FOs, due to their unknown nature, the forcing signals are not included in the power system models used by operation or planning. Therefore, a significant difference in measured and inferred outputs implies an inaccuracy in the assumed model, e.g., due to the forcing signals. The model inference-based methods may differ from each other in terms of the input, output, model, and the use of the model and/or input, e.g., whether a time-domain simulation is needed or not. Although power systems are nonlinear, all methods reviewed below are all based on linearized power system models, and mainly consider FOs in generators. As an example, the hybrid simulation-based method summarized in [7] is also one such method.

There are two model inference-based methods that only require the model of the local generator. These methods are good for distributed deployment because the information about power network and other generators is not needed. The first method is called the effective generator impedance method [54], which derives the generator impedance matrix Z from the generator model, and takes the terminal voltage V as the measured input. The terminal current is the output, which is inferred by $\hat{I} = Z^{-1}V$. Provided the measured output I, if the difference $||\hat{I} - I||_2$ is greater than a pre-specified threshold, then this generator is the FO source. Measurement noises and uncertainties in generator parameters are addressed in [55]. The other method [56] derives a transfer function H between the terminal grid frequency f (the input) and the generator's active power output P_e (the output). Similarly, the measured f and the derived H can infer the output as $\hat{P}_e$. The difference between $\hat{P}_e$ and measured P_e is used for OSL.

Another relevant work in [15] also locates the source of oscillation by assuming the model of local generator is available. The relevant problem considered in [15] is that after the oscillation source is found to be at a specific bus with multiple generators connected, how to further pinpoint the generator that is driving the oscillations. First, the playback simulation is performed with the measured terminal voltage for each generator at identified bus with FO source. Then, dynamic state estimation is applied to the playback simulation results to estimate machine and control states of that generator, and the residue is calculated for that generator. Repeating this process for every generator on that bus and comparing the resulting residues can identify the generator that contains the source of oscillations. As pointed out by the paper, this method is only applicable when there is only one FO happening at a time.

There are several other methods that require the model of the entire system. These methods have to be implemented in a centralized manner. Paper [57] derives the transfer function between generator states (the output) and measurements (the input) assuming the FO source is at one generator. Therefore, if there are N generators, N transfer functions will be derived each with the FO source at one generator. The transfer function corresponding to the smallest normalized error in generator states indicates the source of FO. Extensive numerical studies show that this method works great in many cases, but it does not work under resonance conditions or in situations where FO is located at one of the two or more generators connected to the same bus. The second method [58] only derives a single transfer function between the mode shape at measurement locations (the output) and the vector consisting of all external periodic disturbance injections at generators (the input). The largest entry in the inferred input indicates the location of FO. Tests on Kundur 2-area system and a 68-bus system demonstrate that this method works for both resonance conditions and the situations where there are multiple FO sources. The third method [59,60] directly derives an observer to quantify the external periodic disturbance injections in generator's mechanical power. If an FO occurs in the mechanical power of the ith generator, then only the ith output of the observer shows periodical variation while other outputs of the observer are identically zero. Upon the modeling assumption, this method is only applicable for FOs in machine governor controls. The last method reviewed here, similar to the third method, derives an unknown input observer [12] whose problem formulation considers FO sources in both excitation controls and governor controls. The difference between inferred responses by model and the measured responses is used to establish a one-to-one relationship between a residual and an FO source.

1.1.6   Purely data-driven methods

This subsection summarizes a list of OSL methods that only utilize measurement data, without requiring any explicit information about the system model or topology. These methods are either based on pure data manipulations, e.g., spectrum analysis, matrix operation and matrix decomposition, or assuming that the data are dominated by certain underlying models, e.g., power system classical model, and then estimating the unknown parameters/states to conclude the location of the oscillation source.

The first group of methods assumes certain form of the underlying power system model, but their algorithms are still fully data-driven. It should be noted that these methods can be easily validated and benchmarked in simulation studies when system models are available. The first method concerns natural mode oscillations [61] and assumes a classical modeling of power system, i.e., machines are represented by swing equations and loads by constant impedance. Then, conceptually, Kron reduction can eliminate all non-machine buses such that the resulting system has the same number of machines but with a reduced, fully connected network. Measurement data are used to estimate all unknowns in the Kron reduction network model as well as the unknown damping coefficient of all machines. Finally, a negative sign of the estimated damping coefficient indicates a source of oscillation. This method is tested with a 9-bus system and a 39-bus system. The second method concerns FO and it is based on the sparse identification of nonlinear dynamics (SINDy) [62], which also assumes a classical modeling of power system but with external forces. The first step of this method is to estimate the FO frequency. The estimated frequency is used to constitute the basis functions, i.e., cosine functions, with unknown coefficient matrix in a forced classical power system model. Measurement data are used to estimate the unknown coefficient matrix as well as other unknowns, where there is a parameter dedicated for adjusting the sparsity of the unknown coefficient matrix. Finally, the generator associated with the largest entry of the coefficient matrix is identified as the FO source. This method is tested with the 179-bus system, a 68-bus system, and 6 real-life oscillation events from ISO New England and shows promising results. The third method [63] in this group assumed an additive model comprising the initial state and inputs grouped over time with unknown coefficients. These unknowns are estimated by the group linear absolute shrinkage and selection operator (LASSO) method. This method is tested with a 68-bus system. The last method [64] deals with FOs in the governor system of a generator. This method identifies a transfer function between each machine's electrical power output and its rotor speed using measurement data collected under normal operating conditions. When an FO event occurs, the event data are used, along with the identified transfer function, to calculate residues. The generator that has residues containing a significant sinusoidal component is identified as the FO source. This method is tested with the Kundur 2-area system.

The second group of methods are based on pure data manipulations without explicitly assuming any underlying system model information to be available. Granger causality analysis [65] and spectral Granger causality analysis [16] are introduced for OSL, which calculate the causal influence, either in time-domain or in frequency-domain, between any pair of time series and then such information is used to determine the location of FO. Pearson correlation coefficient (PCC) [66] is used to measure the linear dependency between two time series. If PCC is close to 1 or -1, it means the two time series are strongly correlated. PCC is adopted in [66] to quantify the correlations between an oscillation mode and the device status from Energy Management System data. Devices with $|\text{PCC}| \approx 1$ are possibly the source of oscillations. Robust Principal Component Analysis is introduced in [67] to locate the source of oscillations by decomposing the high-dimensional measurement data matrix into a low-rank matrix and a sparse matrix. The source of oscillations is identified at the location associated with the largest entry in the sparse matrix. This method is tested with a 68-bus system, a 179-bus system and a field event from Electric Reliability Council of Texas. Multi-delay self-coherence spectrum is used for OSL in [68]. The location with signals whose spectra are high, i.e., close to 1, is identified as the source of oscillations.

Another method proposed in [69] requires generator rotor speed and active power output data from all generators in a system. A three-stage selection process based on a set of pre-defined functions gradually screen out the generators and finally point to one generator as the source of oscillations.

1.1.7 Other methods

There are another two OSL methods that can hardly be categorized based on the above classification. In the first method [70], a power system is modeled as an undirected graph, where generator internal buses are vertices, while edge weights are defined by the magnitude of the P component of the energy flow [71]. The top-2 largest eigenvalues of the weighted adjacency matrix can be used to identify the location of FO source. Tests are conducted on a 9-bus system and a 39-bus system. The second method is based on data fusion Dempster–Shafer (D–S) evidence theory [72]. Basically, three OSL methods are synthesized: dissipating energy flow method, OSL based on the phase difference index (time integral of oscillation phase difference of two adjacent bus angles), and OSL based on oscillation phase difference between ΔP_{m} and ΔP_{e}, i.e., deviations of generator mechanical power and electrical power from their steady states. Though the implementation of each individual OSL method may affect their own performance, the hope is that the fusion of multiple OSL methods could lead to a better overall OSL performance. The unknown parameters in the fusion model are solved by minimizing the errors of the three OSL methods over a training data set. Finally, the bus with the largest synthesis decision value is identified as the source of oscillation.

1.1.8   OSL methods for inverter-based resources

As claimed at the beginning of this chapter, the review in this section mainly covers the OSL methods for low-frequency electromechanical oscillations. Still, this subsection is dedicated to a very brief coverage of OSL methods for oscillations involving inverter-based resources (IBRs) that can be drivers/participators of either low-frequency or high-frequency oscillation modes. Similar to energy flow-based OSL methods for low-frequency oscillations, sub-/super-synchronous power flow is proposed to locate the equipment having destabilizing effects in oscillation events involving IBRs [73–82]. AI-based OSL methods are also emerging for IBR oscillations [83,84]. It is worth mentioning that an interesting interpolation FFT algorithm is used in [73] to analyze high-frequency oscillations based on low-speed data from Phasor Measurement Units (PMUs). An interesting mode shape-based method is introduced in [85] which characterizes an IBR's negative damping contribution by the in-phase relationship between oscillations in voltage and reactive power from that IBR, while an out-of-phase relationship implies a positive damping contribution. In addition, it was reported in [86] that a 22-Hz oscillation event involving IBRs was initially identified as an 8 Hz oscillation event by PMU data. This calls for the use of high-speed phasor data or point-on-wave data for analyzing high-frequency oscillations.

1.2   Practical considerations for implementing OSL methods

The majority of the proposed OSL methods are tested in a simulated environment and do not account for specifics of actual PMU measurements or particular power system operating conditions, which may not guarantee the performance of the method and robustness of results in practical settings. An OSL method used as a practical application should be implementable, efficient, and robust for the variety of possible operating conditions and types of oscillations. This section describes some requirements related to the implementation of OSL methods for practical use.

Some OSL methods require PMU measurements from all generators. This requirement is very difficult to satisfy because typically PMUs are installed only at a limited number of generators. Monitoring all generators by PMU could be considered a viable option only at some future stage, with or without a state estimation.

Machine learning-based and model inference-based methods assume the availability of an accurate dynamic model of power systems, generally, at any moment of time. That is not an easy-to-satisfy requirement. Due to multiple reasons, the available dynamic models may have inaccuracies and the level of inaccuracy can vary over time. It is difficult to predict how these unknown inaccuracies can affect the OSL results.

Actual PMU measurements are imperfect because of bad data, missed samples, outliers, and stalled and noisy measurements. Any of these factors, depending on the specific numerical procedures used by the OSL methods, can lead to a failure of the methods or misleading results. For example, a significantly noisy single PMU measurement among many good PMU signals used in the frequency domain analysis could be quite a negatively affecting factor. Another example is a stalled phase value for voltage or current at a non-nominal frequency can result in a false and very high magnitude of low-frequency oscillations in both active and reactive power. Field tests with PMU equipment can also result in other abnormal conditions.

Tripping a transmission element in the middle of an oscillation event, which can be completely unrelated to oscillations, could be harmful to frequency-domain analysis used by OSL methods. Reflection of the tripping event in PMU data looks like a step function leading to a significant distortion of oscillatory spectra. Therefore, it is better to avoid the use of such PMU signals for frequency-domain analysis.

All these factors reinforce the need for PMU pre-processing to filter out any potential issues. It is difficult to envision the need to model all the above situations in a simulated testing environment. All these factors were encountered from a large number of actual oscillatory events during a prolonged analysis of OSL results at ISO New England.

In addition, the sampling rate of PMU measurement as a limiting factor for the highest-frequency components of detected and analyzed oscillations should be taken into consideration. Per Nyquist–Shannon sampling theorem, the theoretically highest frequency of oscillations that can be captured by a sampling rate is half of the rate. For example, that is 15 Hz for a 30 sample-per-second PMU measurement. Practically, the majority of existing PMUs have internal filtering of high-frequency components which, therefore, reduces the theoretical frequency range. For instance, the practically reliably identifiable highest frequency oscillations for PMU with 30 samples per second are somewhere in the range up to 8–12 Hz.

1.3 Studies on dissipating energy flow method

1.3.1 A brief introduction of DEF

DEF method is an implementation of the energy-based OSL method [71], which was first introduced in [2]:

$$
\begin{aligned}
W_{ij}(t) &\approx \int \left(\Delta P_{ij} d\Delta\theta_i + \Delta Q_{ij} \frac{d(\Delta V_i)}{V_i^*} \right) \\
&= \int \left(2\pi \Delta P_{ij} f_i dt + \Delta Q_{ij} \frac{d(\Delta V_i)}{V_i^*} \right) \\
&\approx DE_{ij} t + b_{ij}(\omega t)
\end{aligned}
\tag{1.1}
$$

where the time-variant function $W_{ij}(t)$ and the constant DE_{ij} respectively quantify the dissipating energy and DEF flowing from bus i to bus j; and $b_{ij}(\omega t)$ represents a function containing frequencies at the oscillation frequency and its integer multiples. See more details in [2]. Note that in addition to branches connecting two buses, DEF can also be calculated at the terminal of any devices connected to a single bus.

Similar to the power flow that describes MW/Mvar power defined at 50/60 Hz, DEF is the transient energy flow defined at the oscillation frequency. Out of the total "oscillation energy," DEF represents the monotonically changing component that flows in one direction: from the oscillation source(s) to the oscillation sink(s). DEF can be calculated using synchronized measurements of four electrical quantities, i.e., active power P, reactive power Q, voltage magnitude V and bus angle θ, where the bus angle can be replaced by bus frequency f. Note that active and reactive powers are not directly measured by PMUs, instead, current phasors are measured and then powers can be derived from current and voltage phasors.

According to (1.1), if a power system is under steady state, then $W_{ij}(t) = DE_{ij} = 0$ everywhere. If the power system is experiencing oscillations, then DE_{ij} can become non-zero at different places. Tracing the direction of DEF across the power network can help pinpoint the component(s) causing the oscillations.

Based on the DEF method, an Oscillation Source Locating online application was developed in 2015 and integrated into ISO New England's control room in September 2017, which has been functioning 24/7 since then [87]. It has processed $1,200+$ oscillation events by March 2021 [3,35,36], and it correctly identified the oscillation source in terms of source generator and source area for all instances that can be verified, i.e., having known sources inside and outside ISO New England. The DEF method has also been implemented by the following entities or projects (it should be noted that different implementations of DEF may differ from each other in terms of effectiveness [88]):

- System Operator of the United Power System in Russia, which found that DEF helped them correctly identify the source of oscillations in nine real oscillation events that occurred between 2018 and 2020 [37].
- Transient Security Assessment Tool (TSAT) by PowerTech Labs [89].
- Eastern Interconnection Situational Awareness Monitoring System (ESAMS) Prototype Demonstration Project sponsored by U.S. Department of Energy [42].
- Power Grid Dynamics Analyzer (PGDA) by Electric Power Group [90].

An offline version of the Oscillation Source Locating online application, called OSLp, in MATLAB® has been made available by ISO New England for free upon request [91], which has been shared with several entities throughout the world, including:

- National Renewable Energy Laboratory [92], Southwest Power Pool, Clarkson University in the United States
- Power System Operation Corporation, Indian Institute of Technology, branches in Patna, Indore and Kharagpur, in India
- National Grid Electricity System Operator in the United Kingdom
- Australian Energy Market Operator Limited in Australia

Though DEF has been demonstrated to be effective in practice, it is still not perfect in theory. As mentioned in Section 1.1.3, one fundamental difficulty stems from constructing an energy function for general lossy multi-machine power

Table 1.1 Summary of two implementations of DEF method by ISO New England

DEF implementation	Notes
OSLp V1.3	This is the latest version of OSLp by ISO New England as of September 2022 that can be shared upon request at [91]. Several pre- and post-processing techniques are integrated for better performance when dealing with real PMU data, including automatic data window selection, band pass filtering, spectral analysis, and spikes analysis
OSLp V2.0	This is a beta version of OSLp that is being developed internally within ISO-NE and will be sharable later. This version contains all features in V1.3, and new features include: (1) implementation of the Complex DEF [45], (2) implementation of a cross power spectral density (CPSD) approach [88], (3) classification of oscillation source in terms of active power-frequency control or reactive power-voltage control, (4) enhanced capability for noise filtering, and (5) enhanced data processing to handle PMU data from low-voltage distribution systems potentially containing high-frequency oscillations involving inverter-based resources

systems considering realistic load modeling. Paper [29] provides insights on a complex dependency between generator controls and the dissipating energy which could lead to DEF failure as shown in examples on the Kundur 2-area system (while a DEF failure example on a real event from WECC system is also included in the paper but the results cannot be verified). Paper [30] provides proof based on the passivity theory showing the condition when DEF works or potentially fails.

To evaluate DEF's effectiveness in processing potential challenging cases, larger systems with more realistic settings are tested and reported in this section. In addition to the reported tests [2,36], including 20+ simulated cases on a WECC 179-bus system [93] and 1,000 real events from ISO New England [36], in this section, we further test DEF on a set of longitudinal power systems [3] and multiple cases from IEEE-NASPI OSL Contest [4,94] which are simulated on a 240-bus WECC system with detailed machine modeling plus controls. For all tests in this section, two implementations of DEF by ISO New England are adopted, i.e., OSLp V1.3 and OSLp V2.0. Their features are summarized in Table 1.1. Note that most results are from OSLp V1.3, while OSLp V2.0 will be compared at the end of this section.

1.3.2 Study of idealized longitudinal power systems

Paper [29] provides examples of DEF failure for some FOs in a small Kundur 2-area system. Such a DEF failure is caused by the feedback interaction properties of the system as stated in the paper, which, in other words, is basically the system response to FOs in the oscillation phase and magnitude of each state variable, i.e., oscillation shape of FOs. The oscillation shape of FOs not only depends on the location, frequency, magnitude, and other configurations of FOs but also depends on the mode shape of other system natural modes [95,96]. Such a complex relationship makes it difficult to analytically evaluate the effectiveness of DEF in more practical settings. To this end, idealized longitudinal power systems with known properties are designed and tested in this subsection [3].

It is necessary to emphasize that the source of sustained oscillation can be either the component/control containing the FO, i.e., external periodic disturbance injections, or the badly-tuned/malfunctioning control, causing poorly-damped natural mode oscillations. For FOs, the definition of oscillation source is conceptually well-defined, which is the external periodic disturbance driving the entire system to oscillate. Especially, that external periodic disturbance is not part of the system model used for normal operation and planning. However, for poorly-damped natural mode oscillations, the zero or negative damping of a natural oscillation mode is a system property, determined by the associated eigenvalue of the underlying linearized system model. In general, it is not straightforward to conclude which single or few components/controls are causing this system-level problem, as there could be many components participating in a single mode like an inter-area mode. The left and right eigenvectors can only quantify the magnitude of their participation, but not their contribution, i.e., stabilizing or destabilizing. In some special cases, it can be possible to recognize the damping contribution of a generator. For example, in the simulation environment, setting the damping coefficient D of a generator in classical model to a negative value, or changing the sign of the power system stabilizer's gain of a detailed generator model can naturally define these generators as a source of destabilizing effects.

Paper [29] only considers the oscillation source in the form of FOs, but not the source of the negative damping effect. Therefore, the DEF failure reported in that paper is only limited to the ability of DEF to identify the source of FOs. In this subsection, a number of idealized longitudinal power systems will be studied to demonstrate that *the identification result by the DEF method is either the source of FOs, or the source of the negative damping effect*. Specifically, the following two questions are addressed in the tests:

• How does the system size affect an individual generator's damping contribution to the overall damping of a natural mode under conditions with or without FOs?
• To what extent does the complex interaction introduced in [29] failing the DEF method in the small Kundur system (in terms of locating the source generator of FOs) still remain in large power systems?

1.3.2.1 Tests with only poorly-damped natural mode(s)

A class of idealized longitudinal power systems are considered, whose one-line diagram is shown in Figure 1.1. These systems are selected for testing mainly because (i) system size can be easily adjusted by adding/reducing the identical segments, and (ii) the system does not contain an obvious weak link like in 2-area Kundur system that significantly contributes to the creation of inter-area mode, and (iii) there exists an analytical solution for eigenvalues and eigenvectors [97] such that how the interesting inter-area mode varies with the system size can be easily understood. Other settings about the test systems are listed below. With these settings, the mode shape of the lowest frequency mode in the longitudinal power system with $n = 16$ generators is shown in Figure 1.2. In this 16-machine system,

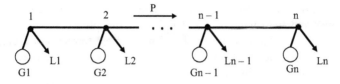

Figure 1.1 One-line diagram of idealized longitudinal power systems

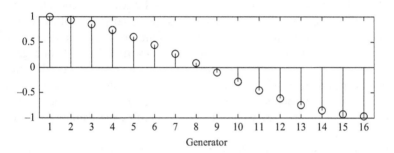

Figure 1.2 Mode shape of the lowest frequency mode in 16-generator system

generators 1–8 form a coherent group oscillate in phase while against another coherent group of generators 9–16, which mimics the wide-spread inter-area oscillation mode in a bulk electric system. The eigenvalue associated with the lowest frequency mode can be approximated by the formulas in (1.2) [97].

- The second-order classical model is used for all generators, where the damping coefficient D can be viewed as an explicit model of the generator's damping contribution: $D > 0$ means positive damping contribution, i.e., stabilizing effects, while $D < 0$ means negative damping contribution, i.e., destabilizing effects.
- Unless otherwise specified, all generator parameters are identical: $H = 3.0$ and $D = 4.0$ in pu.
- All bus voltage magnitudes are regulated to be 1.0 pu.
- All loads, except for L_1 and L_n, are identical: $L_k = 50$ MW $+ j20$ Mvar for any $k \in \{2, 3, \cdots, n - 1\}$, while $L_1 = 20$ MW $+ j20$ Mvar and $L_n = 70$ MW$+ j20$ Mvar, creating a $P = 30$ MW active power transfer from Bus 1 to Bus n. All loads are represented by constant power model.
- All lines are identical with impedance parameters: $R = 0.0$ and $X = 0.5$ in pu.
- With the constant power load model and lossless network, the source/sink of oscillations can only be generators:

$$
\begin{aligned}
\lambda &= a + j\omega \\
\omega &\approx \frac{\pi\sqrt{K_c}}{N} \\
a &\approx -\frac{K_\omega}{2} - \frac{K_d\pi^2}{2N^2}
\end{aligned}
\tag{1.2}
$$

where K_c is the synchronizing torque coefficient, K_d is the damping coefficient due to the damper and field windings, and K_ω is the damping coefficient due to the frequency dependence of the electrical and mechanical torque. For a classic generator model in the idealized longitudinal system, damping coefficient D corresponds to K_ω and $K_d = 0$ in (1.2).

In the first test case, a 4-machine system is considered where the negative damping effect is introduced at generator 1 by setting $D_{G1} < 0$, while keeping $D = 4.0$ for the other three generators. The lowest frequency mode of the system is around 0.87 Hz, whose damping is shown in Table 1.2 with different values for D_{G1}.

From this simple test case, two useful and potentially generalizable observations are made: **Observation 1:** Damping of a natural mode is the total effect of the damping contribution from all generators; **Observation 2:** With one generator having a negative damping effect, the total damping of the system oscillation mode can still be positive, which is a result of other generators providing a greater, positive damping contribution. These observations can be supported by the DEF result, shown in Figure 1.3, which directly quantifies the damping contribution from each generator.

The second test case considers $n = 10$ generators. In this case, two modes with the lowest frequencies, i.e., 0.43 Hz and 0.78 Hz, become poorly damped when D_{G1} takes a very negative value, i.e., -13.0, as shown in Table 1.2. For each of the two modes, DEF method can be applied using the simulated system responses to

Table 1.2   *Eigenvalue and damping of 0.87 Hz mode vs. damping coefficient D_{G1} in 4-machine system*

D_{G1}	Eigenvalue	Damping
4.0	$-0.333 + j5.46$	6.1%
-2.0	$-0.09 + j5.48$	1.6%
-4.0	$-0.006 + j5.48$	0.1%

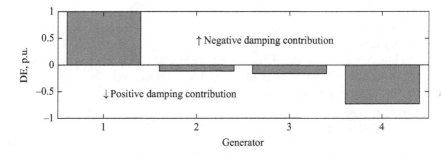

Figure 1.3   *Damping contribution quantified by DEF in 4-machine system with*
$D_{G1} = -4.0$

Table 1.3   Eigenvalue and damping of 0.87 Hz and 0.43 Hz modes vs. damping coefficient D_{G1} in 10-machine system

D_{G1}	Eigenvalue	Damping
4.0	$-0.33 + j2.74$	12%
	$-0.33 + j4.91$	6.8%
-4.0	$-0.19 + j2.77$	6.8%
	$-0.19 + j4.92$	3.9%
-13.0	$-0.017 + j2.80$	0.6%
	$-0.006 + j4.95$	0.1%

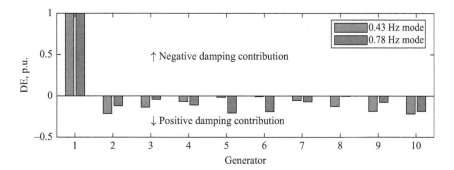

Figure 1.4   Damping contribution quantified by DEF in 10-machine system with $D_{G1} = -13.0$

quantify each generator's damping contribution to that mode. DEF results in Figure 1.4 shows that generator 1 has a large negative damping contribution to both of the two modes, while all other generators have a positive damping contribution.

Comparing the above two test cases, it can be seen that to destabilize the low-frequency modes in a larger power system, a stronger negative damping effect is needed at a single generator. In addition, these two cases illustrate that the DEF method can identify the source of the negative damping effect in both small and large longitudinal power systems. This conclusion also holds true for meshed power systems, which has been demonstrated in [2] on a 179-bus WECC system model.

1.3.2.2   Tests with both poorly damped natural mode(s) and FOs

FOs are added to the longitudinal power systems in Figure 1.1 to evaluate how the negative damping effect could affect the DEF's ability to identify the source of FOs. Two groups of tests are considered with different modeling of the negative damping effect. In the first group, classical generator model is used where the damping coefficient of one generator is set to a negative value to introduce a

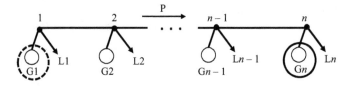

Figure 1.5   One-line diagram of idealized longitudinal power systems with FOs

negative damping effect. In the second group, a negative damping effect is introduced by setting the gain of the power system stabilizer to a negative value at one generator.

The one-line diagram of the test system is shown in Figure 1.5, where the negative damping effect, if any, is added to generator 1, while the FO, if any, is added to generator n. All other settings remain unchanged as in the system in Figure 1.1.

In **the first group of tests**, the forced signal is added to the mechanical power at generator n in the form specified in (1.3), which represents a 0.01 pu variation of mechanical power input around its steady state. The frequency f is selected to be very close to the lowest frequency of all natural modes, e.g., 0.87 Hz for $n = 4$, 0.43 Hz for $n = 10$ and 0.28 Hz for $n = 16$. Such a selection of f is made on purpose to cause resonance conditions for a system-wide natural mode. In addition, the damping coefficient D_{G1} is set to a negative value. Detailed results are shown in Figure 1.6(a):

$$P_m = P_0 + 0.01 \sin (2\pi f t) \tag{1.3}$$

In the 4-machine system, let D_{G1} take -0.1, -1.0, and -2.0, respectively, and apply DEF to quantify the damping contribution of each generator. As shown in the first plot of Figure 1.6(a), for a moderate negative damping contribution from G1 with $D_{G1} = -0.1$, DEF finds G4 to have the largest dissipating energy injection. Therefore, DEF has no issues identifying the FO source G4. However, for a large negative damping contribution from G1, DEF finds that G4 either injects dissipating energy (when $D_{G1} = -1.0$) that is less than that of G1, or absorbs dissipating energy (when $D_{G1} < -1.0$) misleadingly implying that G4 is an oscillation sink that stabilizes the system. In these cases, DEF can falsely claim G1 as the source of FOs.

The second plot of Figure 1.6a illustrates the DEF results in the 10-machine system. For a wide range of negative damping contributions from G1 with $D_{G1} > -6.0$, DEF always finds G10 to have the largest dissipating energy injection such that DEF has no issues identifying the FO source at G10. Only when the negative damping contribution from G1 becomes very large, say $D_{G1} < -6.0$, should the DEF falsely claim G1 as the FO source.

This trend continues into the 16-machine system with FOs, where DEF always finds G16 as the largest FO source unless an unrealistically large negative damping is produced by G1 say $D_{G1} < -13.0$. It should be noted that those values of

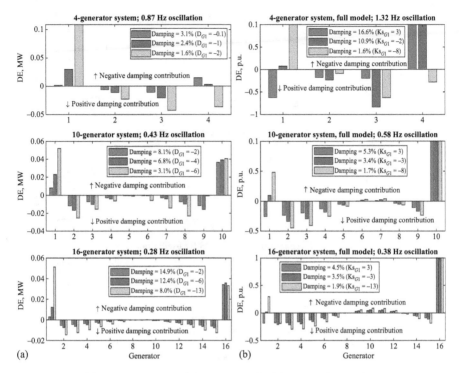

Figure 1.6   (a) Damping contribution quantified by DEF for each generator in systems with both FOs (added to G4, G10, and G16, respectively) and negative damping effect (added to G1 with negative D_G). (b) Damping contribution quantified by DEF for each generator in systems with both FOs (added to G4, G10, and G16, respectively) and negative damping effect (added to G1 with negative gain in its power system stabilizer)

$D_{G1} < -13.0$ are unpractical because a number of unstable modes would be created.

The second group of tests is almost the same as the first group, but with more detailed modeling of machines and their controls, and the realization of the negative damping effect at G1: generators are represented by GENROU, exciters by SEXS, power system stabilizers by IEEEST, and governors by TGOV1. The negative damping effect at G1 is achieved by setting the gain of G1's power system stabilizer to a negative value. Note that using the detailed machine model with controls changes the oscillation frequency of the corresponding modes. The three modes in Figure 1.6(a) respectively become 1.32 Hz, 0.58 Hz, and 0.38 Hz. The DEF results are shown in Figure 1.6(b), which are consistent with those of the first group. Specifically, in the 10-machine and 16-machine systems, DEF has no issues identifying the FO source at G10 or G16 with any reasonable gain value for G1's power system stabilizer.

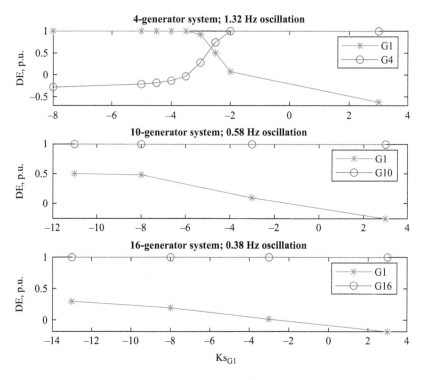

Figure 1.7   Damping contribution quantified by DEF for G1 (negative damping effect) and Gn (FO source) considering different sizes of negative damping effect from G1

In addition, DEF results for the second group of tests with different gain values for G1's power system stabilizer are summarized in Figure 1.7. A few important observations are drawn below.

- False FO source identification by DEF occurs only for the 4-machine system and only for the high negative damping effect from G1.
- Oscillation source identified by DEF, i.e., largest dissipating energy injection, is either the source of FOs, or the source of negative damping effect. The ability to identify both the source of FO and the source of negative damping is an important property of DEF for practical use.

1.3.3   Study of multiple cases from IEEE-NASPI OSL contest

In this subsection, the DEF method is tested on a meshed power system with more realistic settings using hundreds of challenging cases to better show its effectiveness and limitation. The test system is the reduced WECC 240-bus system developed by National Renewable Energy Laboratory, whose one-line diagram is shown in Figure 1.8 [98]. This test system has 4 areas, 243 buses, 139 loads, and 146 generating units at 56 power plants, including 109 synchronous machines and

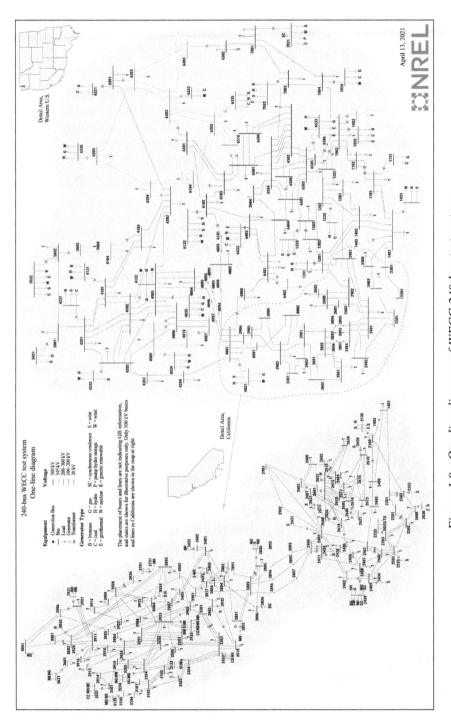

Figure 1.8 One-line diagram of WECC 240-bus test system

37 renewable units. Each generator is represented by the detailed model with exciter and governor, and each load by a ZIP load model with 40%, 30%, and 30% of constant impedance, constant current, and constant power components, respectively. This system is used for IEEE-NASPI OSL Contest [4] and its data is publicly available at [94].

FOs are added to every generator, one per each study case, either to the SEXS excitation system at each of the 109 generators, or to the TGOV1 governor system at each of the 37 generators with TGOV1 governor. Resonance conditions are intentionally introduced respectively at three system natural modes to make the test cases more challenging: the lowest frequency inter-area mode at 0.379 Hz, an inter-area mode at 0.614 Hz, and a local mode at 1.27 Hz. All generators are made observable by PMU measurements in this test.

Three criteria used for quantifying DEF's success rate are:

1. Identification of a **generator** containing the FO source. The generator with the largest dissipating energy injection is identified by DEF as the FO source.
2. Identification of a **power plant** containing the FO source. Note that in this test system, more than one generator is installed at the majority of power plants. The power plant with the largest dissipating energy injection at the point of interconnection is identified by DEF as the FO source.
3. Identification of an **area** containing the FO source. The area with the largest dissipating energy going outward (a sum of dissipating energy on all tie-lines) is identified by DEF as the FO source.

The results are summarized in Table 1.4. An overall observation is that the DEF method correctly identifies the FO source in the governor for all test cases, while it can potentially fail to locate the FO source if the FO source is located in an exciter. Recall that the DEF method has a very high success rate for practical FO events [36], this is mainly because of the fact that the vast majority of practically observed FOs have the source in the active power control of traditional generators. However, for FO sources in exciter systems, the DEF method may fail, especially for low-frequency oscillations. For example, when FOs at 0.379 Hz are respectively added to 93 generators to cause a resonance condition with the lowest frequency inter-area mode of the system, the DEF method can only identify the FO source at the correct location for 12 cases, i.e., a very low success rate at 12.9%. Typically, DEF identifies another generator at the same power plant as the source. The DEF method can maintain a relatively high success rate for FOs resonating with local modes, and for identification of source power plant or source area. It is worth mentioning that the tested cases are all challenging resonance cases, while real-life FO events are not always involving a resonance condition.

The low success rates of DEF for FO sources in exciter systems for simulated cases are concerning, as such scenarios, although very rare in practice, are still possible and should be taken into consideration. The actual oscillation events, however, have not confirmed such a concern so far. For example, ISO New England had only few confirmed events of FO originating from the excitation systems during 5 years and the online OSL application has correctly identified the

Table 1.4 *DEF results on the 240-bus WECC test system*

Location of FO	Frequency of FO Hz	# of tested FO cases	Identification of source generators			Identification of source power plants			Identification of source areas		
			# of cases that DEF succeeds	Success rate	Average success rate	# of cases that DEF succeeds	Success rate	Average success rate	# of cases that DEF succeeds	Success rate	Average success rate
Governor	0.379	31	31	100%	100%	31	100%	100%	31	100%	100%
	0.614	31	31	100%		31	100%		31	100%	
	1.270	31	31	100%		31	100%		31	100%	
Exciter	0.379	93	12	12.9%	60.1%	79	84.9%	86.9%	83	89.2%	94.7%
	0.614	96	71	74.0%		76	79.2%		91	94.8%	
	1.270	91	85	93.4%		88	96.7%		91	100%	

generator containing FO source for all these events. Still, to better handle FO sources in exciter systems, the CDEF method [45] is implemented and tested for all FO cases listed in Table 1.4 for frequencies 0.379 Hz and 0.614 Hz. A correct value of the user-selected parameter in CDEF, i.e., the projection coefficient K, is difficult to identify because that parameter reflects the overall characteristics of all system components including transmission lines, transformers, and loads. In this test, the coefficient K has been varied from 0 to 1 with a step of 0.05 and the results are summarized in Figure 1.9. This figure clearly shows an improved OSL success rate with an increase of K for FOs in exciter systems, and a decreased OSL success rate for FOs in governor systems. Such an observation is consistent with [45]. Notably, the success rate for FOs in exciters in terms of identifying the correct generator with a frequency at the system's lowest frequency mode becomes >60% with $K = 0.2$, where the success rate for FOs in governors remains at close to 100%. Therefore, $K = 0.2$ seems the right selection for this reduced 240-bus WECC system. For real-world power systems, the optimal selection of K in the CDEF method is still an open question mainly because of uncertain load characteristics.

The different impacts on CDEF for FOs in exciters and governors illustrated in Figure 1.9 imply a useful question to answer: whether the FO is in the exciter or in the governor? If the answer to this question can be obtained and fed to CDEF, then, as can be concluded from results in Figure 1.9, the parameter K can be intentionally selected to better handle the FO event under study: $K = 0$ for FOs in governors, while $K > 0.2$ for FOs in exciters. Even for an optimal selection of K, the CDEF method does not guarantee to provide a 100% success rate for FO located in the excitation systems. The likely reason for that is a complex interaction of controls of closely located generators [29].

In addition to CDEF, the CPSD method proposed by Rensselaer Polytechnic Institute [88] is also tested with the cases in Table 1.4. Similar to DEF, the CPSD method provides a 100% success rate for FOs in the governor. However, it is less

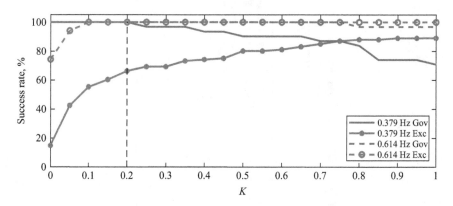

Figure 1.9   CDEF with various projection coefficient K

efficient than CDEF (with a reasonable selection of parameter $K = 0.2$) for the FO sources in the excitation systems: 43% for 0.379 Hz and 84% Hz for 0.614 Hz.

1.4   Conclusion and future work

Sustained oscillations in power systems can be either caused by FOs, i.e., external periodic disturbance injection, or by a certain poorly-tuned controller causing poor damping in a natural oscillation mode. Both the components containing the FO source and the components causing a negative damping effect should be intuitively considered as the source of oscillation. However, there is currently a lack of formal definition of oscillation source for natural mode oscillations.

In this chapter, the recent developments in OSL techniques were reviewed and summarized. Then, the DEF method was tested with several longitudinal test power systems and a reduced WECC 240-bus system to show its effectiveness and limitation. Specifically, based on many numerical experiments, it has been shown that DEF can be effective in dealing with FOs stemming from the generator's governor systems, while may fail for FOs in the generator's excitation systems. Still, by using a series of longitudinal test systems as an example, it was demonstrated that the system component identified by DEF is either the actual FO source or the source of the negative damping effect.

To locate the source of FOs, the recently proposed complex dissipating energy flow method and the cross power spectral density method were also tested and shown to improve DEF's success rate, but not to 100%, yet. With the latest advancements, the OSL problem is still not completely solved. More investigations are still needed for a better solution along with the pursuit of a formal definition of oscillation source for natural mode oscillations.

With more and more IBRs being integrated into modern power grids, oscillations with frequencies higher than the traditional electromechanical frequency range (0.1 to $\sim$ 3 Hz), such as 3–8 Hz and much higher frequencies, are also emerging [92,99,100]. These IBR-related oscillations look more like natural mode oscillations. To deal with these emerging problems, both the measurement system with an adequate sampling rate and OSL methods deserve a revisit.

References

[1]   NASPI. Actual and potential phasor data applications; 2009. https://www. naspi.org/sites/default/files/reference_documents/67.pdf.

[2]   Maslennikov S, Wang B, and Litvinov E. Dissipating energy flow method for locating the source of sustained oscillations. *International Journal of Electrical Power & Energy Systems*. 2017;88:55–62.

[3]   Maslennikov S. Efficiency of the DEF Method for locating the Source of Oscillation; 2021. https://www.wecc.org/Administrative/06_Maslennikov_DEF_Efficiency_for_source_locating_March 2021.pdf.

[4]   Maslennikov S and Wang B. Creation of simulated test cases for the oscillation source location contest: preprint. In: *IEEE PESGM*, 2022.

[5]   Nayfeh AH. *Nonlinear Oscillations*. New York, NY: Wiley-VCH; 1995.

[6]   Wu D, Vorobev P, Chevalier SC, *et al.* Modulated oscillations of synchronous machine nonlinear dynamics with saturation. *IEEE Transactions on Power Systems*. 2020;35(4):2915–2925.

[7]   Wang B and Sun K. Location methods of oscillation sources in power systems: a survey. *Journal of Modern Power Systems and Clean Energy*. 2017; 5(2):151–159.

[8]   Shu Y, Zhou X, and Li W. Analysis of low frequency oscillation and source location in power systems. *CSEE Journal of Power and Energy Systems*. 2018;4(1):58–66.

[9]   Feng S, Zheng B, Jiang P, *et al.* A Two-level forced oscillations source location method based on phasor and energy analysis. *IEEE Access*. 2018;6:44318–44327.

[10] Cai X, Shu Z, Deng J, *et al.* Location and identification method for low frequency oscillation source considering control devices of generator. In: *5th ICISCE*, 2018. p. 818–827.

[11] Guo S, Ouyang F, and Lv F. A new generator control device level disturbance source location method of power oscillation based on PMU. In: *2019 IEEE 3rd Conference on Energy Internet and Energy System Integration (EI2)*, 2019. p. 1558–1562.

[12] Luan M, Gan D, Wang Z, *et al.* Application of unknown input observers to locate forced oscillation source. *International Transactions on Electrical Energy Systems*. 2019;29(9):e12050.

[13] Guo1 S, Zhang S, Li L, *et al.* An oscillation energy calculation method suitable for the disturbance source location of generator control systems. *Journal of Physics: Conference Series*. 2020;1518:1–8.

[14] Luan M, Li S, Gan D, *et al.* Frequency domain approaches to locate forced oscillation source to control device. *International Journal of Electrical Power & Energy Systems*. 2020;117:105704.

[15] Marchi P, Estevez PG, Galarza C. Location method for forced oscillation sources caused by synchronous generators. arXiv:211002692v1; 2021.

[16] Luan M, Li S, and Gan D. Locating forced oscillation source using granger causality analysis and delay estimation. In: *2019 4th International Conference on Intelligent Green Building and Smart Grid*, 2019. p. 502–507.

[17] Wang W, Chen C, Zhu L, *et al.* Model-less source location for forced oscillation based on synchrophasor and moving fast Fourier transformation. In: *2020 IEEE PES Innovative Smart Grid Technologies Europe (ISGT-Europe)*, 2020. p. 404–408.

[18] Banna HU, Solanki SK, and Solanki J. Data-driven disturbance source identification for power system oscillations using credibility search ensemble learning. *IET Smart Grid*. 2019;2(2):293–300.

[19] Zhou N, Ghorbaniparvar M, and Akhlaghi S. Locating sources of forced oscillations using transfer functions. In: *2017 IEEE Power and Energy Conference at Illinois (PECI)*, 2017. p. 1–8.

[20] O'Brien J, Wu T, Venkatasubramanian VM, *et al.* Source location of forced oscillations using synchrophasor and SCADA data. In: *HICSS*, 2017. p. 3173–3182.

[21] Zhang H, Ning J, Yuan H, *et al.* Implementing online oscillation monitoring and forced oscillation source locating at peak reliability. In: *2019 North American Power Symposium (NAPS)*, 2019. p. 1–6.

[22] Zuhaib M, Rihan M, and Saeed MT. A novel method for locating the source of sustained oscillation in power system using synchrophasors data. *Protection and Control of Modern Power Systems*. 2020;5(30):1–12.

[23] Xu Y, Gu Z, and Sun K. Location and mechanism analysis of oscillation source in power plant. *IEEE Access*. 2020;8:97452–97461.

[24] Ortega A and Milano F. Source location of forced oscillations based on bus frequency measurements. In: *2021 IEEE 30th International Symposium on Industrial Electronics (ISIE)*, 2021. p. 01–06.

[25] Roy S, Ju W, Nayak N, *et al.* Localizing power-grid forced oscillations based on harmonic analysis of synchrophasor data. In: *2021 55th Annual Conference on Information Sciences and Systems (CISS)*, 2021. p. 1–5.

[26] Moon YH, Cho BH, Lee YH, *et al.* Derivation of energy conservation law by complex line integral for the direct energy method of power system stability. In: *Proceedings of the 38th IEEE Conference on Decision and Control (Cat. No.99CH36304)*. 1999, vol. 5, p. 4662–4667.

[27] Chiang HD. Study of the existence of energy functions for power systems with losses. *IEEE Transactions on Circuits and Systems*. 1989;36(11):1423–1429.

[28] Chen L, Xu F, Min Y, *et al.* Transient energy dissipation of resistances and its effect on power system damping. *International Journal of Electrical Power & Energy Systems*. 2017;91:201–208. Available from: https://www.sciencedirect.com/science/article/pii/S0142061517300558.

[29] Zhi Y and Venkatasubramanian V. Analysis of energy flow method for oscillation source location. *IEEE Transactions on Power Systems*. 2021;36(2):1338–1349.

[30] Chevalier S, Vorobev P, and Turitsyn K. A passivity enforcement technique for forced oscillation source location. arXiv:190605169v1; 2019.

[31] Jha R and Senroy N. Locating source of oscillation using mode shape and oscillating energy flow techniques. In: *2018 IEEE 8th Power India International Conference (PIICON)*, 2018. p. 1–6.

[32] Jha R and Senroy N. Locating the source of forced oscillation in power systems using system oscillating energy. In: *2018 8th IEEE India International Conference on Power Electronics (IICPE)*, 2018. p. 1–6.

[33] Li X, Zhou M, and Luo Y. A disturbance source location method on the low frequency oscillation with time-varying steady-state points. *CES Transactions on Electrical Machines and Systems*. 2018;2(2):226–231.

[34] Naderi K, Hesami Naghshbandy A, and Annakkage U. A new phase-driven approach to pinpoint source of forced oscillations based on fundamental frequency. *Electrical Engineering*, 2022;104(3):3015–3025

[35] Maslennikov S and Litvinov E. Online oscillations management at ISO New England. In: *Power System Grid Operation Using Synchrophasor Technology*. 2018;257–284.

[36] Maslennikov S and Litvinov E. ISO New England experience in locating the source of oscillations online. *IEEE Transactions on Power Systems*. 2021; 36(1):495–503.

[37] Danilov MA, Rodionov AV, Butin KP, *et al.* Practical results of solving the problem of detecting a source of low-frequency oscillations in the power system by the dissipating energy flow method. In: *2021 4th International Youth Scientific and Technical Conference on Relay Protection and Automation (RPA)*, 2021. p. 1–10.

[38] Yu D. Disturbance source location method of low-frequency oscillation considering information loss of key nodes. In: *2017 IEEE Conference on Energy Internet and Energy System Integration (EI2)*, 2017. p. 1–6.

[39] Popov AI, Dubinin DM, Mokeev AV, *et al.* Examples of processing low-frequency oscillations in Russia and ways to improve the analysis. In: *2022 International Conference on Smart Grid Synchronized Measurements and Analytics (SGSMA)*, 2022. p. 1–6.

[40] Kirihara K, Yamazaki J, Chongfuangprinya P, *et al.* Speeding up the dissipating energy flow based oscillation source detection. In: *International Conference on SGSMA*, 2019. p. 1–8.

[41] Singh I, Reddy Chiluka VK, Trudnowski DJ, *et al.* A strategy for oscillation source location using closed-contour grouping and energy-flow spectra. In: *IEEE/PES T&D Conference and Exposition*, 2020. p. 1–5.

[42] Follum J, Yin T, and Betzsold N. Regional oscillation source localization – Implementation in the ESAMS tool. PNNL Technical Report 29612; 2020.

[43] Wu X, Chen X, Shahidehpour M, *et al.* Distributed cooperative scheme for forced oscillation location identification in power systems. *IEEE Transactions on Power Systems.* 2020;35(1):374–384.

[44] Xu B, Cao Y, Zhang H, *et al.* Localization approach for forced oscillation source based on synchrosqueezed wavelet. In: *2022 IEEE International Conference on Electrical Engineering, Big Data and Algorithms (EEBDA)*, 2022. p. 11–15.

[45] Estevez PG, Marchi P, Galarza C, *et al.* Complex dissipating energy flow method for forced oscillation source location. *IEEE Transactions on Power Systems.* 2022;37(5):4141–4144.

[46] Hu Y, Bu S, Yi S, *et al.* A novel energy flow analysis and its connection with modal analysis for investigating electromechanical oscillations in multi-machine power systems. *IEEE Transactions on Power Systems.* 2022; 37(2):1139–1150.

[47] Jha R and Senroy N. Forced oscillation source location in power systems using system dissipating energy. *IET Smart Grid.* 2019;2(4):514–521.

[48] Banna HU, Solanki SK, and Solanki J. Data-driven disturbance source identification for power system oscillations using credibility search ensemble learning. *IET Smart Grid.* 2019;2(2):293–300.

[49] Talukder S, Liu S, Wang H, *et al.* Low-frequency forced oscillation source location for bulk power systems: a deep learning approach. In: *2021 IEEE International Conference on Systems, Man, and Cybernetics (SMC)*, 2021. p. 3499–3404.

[50] Meng Y, Yu Z, Lu N, *et al.* Time series classification for locating forced oscillation sources. *IEEE Transactions on Smart Grid.* 2021;12(2):1712–1721.

[51] Feng S, Chen J, Ye Y, *et al.* A two-stage deep transfer learning for localisation of forced oscillations disturbance source. *International Journal of Electrical Power & Energy Systems.* 2022;135:107577.

[52] Zheng G, Wang H, Liu S, *et al.* 2021 IEEE-NASPI oscillation source location contest: team woodpecker. In: *IEEE PESGM*, 2022. p. 1–5.

[53] Zheng G. GE's solution – 2021 IEEE-NASPI oscillation source location contest. WECC; 2021. https://www.wecc.org/Administrative/Zheng_GE Solution_Sept 2021.pdf.

[54] Chevalier S, Vorobev P, and Turitsyn K. Using effective generator impedance for forced oscillation source location. In: *2019 IEEE Power & Energy Society General Meeting (PESGM)*, 2019. p. 1–1.

[55] Chevalier S, Vorobev P, and Turitsyn K. A Bayesian approach to forced oscillation source location given uncertain generator parameters. *IEEE Transactions on Power Systems.* 2019;34(2):1641–1649.

[56] Duong TD and D'Arco S. Locating generators causing forced oscillations based on system identification techniques. In: *2020 IEEE PES Innovative Smart Grid Technologies Europe (ISGT-Europe)*, 2020. p. 191–195.

[57] Agrawal U, Pierre JW, Follum J, *et al.* Locating the source of forced oscillations using PMU measurements and system model information. In: *2017 IEEE Power & Energy Society General Meeting*, 2017. p. 1–5.

[58] Cabrera IR, Wang B, and Sun K. A method to locate the source of forced oscillations based on linearized model and system measurements. In: *2017 IEEE Power & Energy Society General Meeting*, 2017. p. 1–5.

[59] Li S, Luan M, Gan D, *et al.* A model-based decoupling observer to locate forced oscillation sources in mechanical power. *International Journal of Electrical Power & Energy Systems.* 2018;103:127–135.

[60] Li S, Gan D, and Luan M. A novel method for locating forced oscillation source in mechanical power based on power spectral density. *Power System Technology.* 2019;43:236–243.

[61] Ping Z, Li X, He W, *et al.* Sparse learning of network-reduced models for locating low frequency oscillations in power systems. *Applied Energy.* 2020;262:114541.

[62] Cai Y, Wang X, Joos G, *et al.* An online data-driven method to locate forced oscillation sources from power plants based on sparse identification of nonlinear dynamics. *IEEE Transactions on Power Systems.* 2022;1–15, https://ieeexplore.ieee.org/document/9822991.

[63] Anguluri R, Sankar L, and Kosut O. Localization and estimation of unknown forced inputs: a group LASSO approach. arXiv:220107907; 2022.

[64] Jakobsen SH, Bombois X, and D'Arco S. Data-based model validation for locating the source of forced oscillations due to power plant governors. In: *5th International Conference on Smart Energy Systems and Technologies*, 2022. p. 1–6.

[65] Li W, Huang T, Frerisy NM, *et al.* Data-driven localization of forced oscillations in power systems. In: *2019 IEEE Innovative Smart Grid Technologies - Asia (ISGTAsia)*, 2019. p. 239–243.

[66] Zuo J, Shen Y, Chen D, *et al.* Low frequency oscillation mode source identification with wide-area measurement system. In: *2019 IEEE 3rd Conference on Energy Internet and Energy System Integration (EI2)*, 2019. p. 1525–1539.

[67] Huang T, Freris NM, Kumar PR, *et al.* A synchrophasor data-driven method for forced oscillation localization under resonance conditions. *IEEE Transactions on Power Systems.* 2020;35(5):3927–3939.

[68] Ghorbaniparvar F and Sangrody H. PMU application for locating the source of forced oscillations in smart grids. In: *2018 IEEE Power and Energy Conference at Illinois (PECI)*, 2018. p. 1–5.

[69] Naderi K, Naghshbandy AH, and Annakkage UD. Three-stage data-driven phase analysis to reveal generator-site origin source of forced oscillations under resonance. *IEEE Access.* 2022;10:62365–62376.

[70] Naghshbandy AH, Naderi K, and Annakkage UD. A Laplacian approach to locate source of forced oscillations under resonance conditions based on energy-driven multilateral interactive pattern. *Iranian Journal of Electrical and Electronic Engineering.* 2022;18(3):1–16.

[71] Chen L, Min Y, and Hu W. An energy-based method for location of power system oscillation source. *IEEE Transactions on Power Systems.* 2013; 28(2):828–836.

[72] Gu J, Xie D, Gu C, *et al.* Location of low-frequency oscillation sources using improved D-S evidence theory. *International Journal of Electrical Power & Energy Systems.* 2021;125:106444.

[73] Xiuping S, Qian L, Shijie J, *et al.* The analysis system of online monitoring and disturbance source locating for subsynchronous oscillation based on D5000 platform. In: *2019 IEEE 2nd International Conference on Electronics Technology (ICET)*, 2019. p. 515–519.

[74] Lei J, Shi H, Jiang P, *et al.* An accurate forced oscillation location and participation assessment method for DFIG wind turbine. *IEEE Access.* 2019;7:130505–130514.

[75] Ma J, Zhao D, Shen Y, *et al.* Research on positioning method of low frequency oscillating source in DFIG-integrated system with virtual inertia control. *IEEE Transactions on Sustainable Energy.* 2020;11(3):1693–1706.

[76] Ma J, Zhang Y, Shen Y, *et al.* Equipment-level locating of low frequency oscillating source in power system with DFIG integration based on dynamic energy flow. *IEEE Transactions on Power Systems.* 2020;35(5):3433–3447.

[77] Xie X, Zhan Y, Shair J, *et al.* Identifying the source of subsynchronous control interaction via wide-area monitoring of sub/super-synchronous power flows. *IEEE Transactions on Power Delivery.* 2020;35(5):2177–2185.

[78] Gao B, Wang Y, Xu W, *et al.* Identifying and ranking sources of SSR based on the concept of subsynchronous power. *IEEE Transactions on Power Delivery.* 2020;35(1):258–268.

[79] Ma Y, Huang Q, Cai D, *et al.* Application of sub/super-synchronous power for identifying the source of subsynchronous control interaction. In: *2021 IEEE Power & Energy Society General Meeting (PESGM),* 2021. p. 1–5.

[80] Wang S and Yang D. Fast oscillation source location method based on instantaneous active/reactive power direction. In: *2021 IEEE 22nd Workshop on Control and Modelling of Power Electronics (COMPEL),* 2021. p. 1–6.

[81] Cai Y, Wang X, Joos G, *et al.* Application of energy flow analysis in investigating machine-side oscillations of full converter-based wind generation systems. In: *IET Renewable Power Generations,* New York, NY: Wiley, 2022. p. 900–901.

[82] Xi X, Xing C, Li S, *et al.* Identification of the oscillation source in multiple grid-connected converters systems. In: *2021 International Conference on Power System Technology (POWERCON),* 2021. p. 182–187.

[83] Du W, Chen J, Wang Y, *et al.* Measurement-driven source tracing of torsional subsynchronous oscillations caused by open-loop modal resonance. *IEEE Transactions on Instrumentation and Measurement.* 2022;71:1–14.

[84] Ren B, Li Q, Wang C, *et al.* A tracing approach for a subsynchronous oscillation source in a power system with grid-connected PMSG. In: *2021 IEEE Sustainable Power and Energy Conference,* 2021. p. 197–202.

[85] Jalali A, Badrzadeh B, Lu J, *et al.* System strength challenges and solutions developed for a remote area of Australian power system with high penetration of inverter-based resources. *CIGRE Science and Engineering.* 2021;20:27–37.

[86] Wang C, Mishra C, Jones KD, *et al.* Identifying oscillations injected by inverter-based solar energy sources. In: *IEEE PESGM,* 2022. p. 1–5.

[87] ISO NEWSWIRE. ISO-NE has developed a groundbreaking software solution to locate the source of dangerous power system oscillations; 2019. https://isonewswire.com/2019/09/06/iso-ne-has-developed-a-groundbreaking-software-solution-to-locate-the-source-of-dangerous-power-system-oscillations.

[88] Osipov D, Konstantinopoulos S, and Chow JH A cross-power spectral density method for locating oscillation sources using synchrophasor measurements. In: *IEEE Transactions on Power Systems,* Early Access, Dec. 2022.

[89] PowerTech Labs. TSAT 20.0 Release Notes; 2020. https://www.dsatools.com/wp-content/uploads/2020/05/TSAT-20.0-Release-Notes.pdf.

[90] Nayak N. EPG products update. In: *JSIS Meeting,* 2019.

[91] ISO New England. Oscillation Source Locating Software. https://www.iso-ne.com/participate/support/request-software/.

[92] Dong S, Wang B, Tan J, *et al*. Analysis of November 21, 2021, Kauai Island Power System 18–20 Hz Oscillations; 2022.

[93] Maslennikov S, Wang B, Zhang Q, *et al*. A test cases library for methods locating the sources of sustained oscillations. *In: IEEE PESGM*, 2016. p. 1–5.

[94] IEEE-NASPI Oscillation Source Location Contest. https://www.naspi.org/node/890.

[95] Myers RB and Trudnowski DJ. Effects of forced oscillations on spectral-based mode-shape estimation. In: *IEEE PESGM*, 2013. p. 1–6.

[96] Zhi Y and Venkatasubramanian V. Interaction of forced oscillation with multiple system modes. *IEEE Transactions on Power Systems*. 2021; 36(1):518–520.

[97] Ustinov SM, Milanovic JV, and Maslennikov SA. Inherent dynamic properties of interconnected power systems. *International Journal of Electrical Power & Energy Systems*. 2002;24:371–378.

[98] NREL. Test case repository for high renewable study; 2019. https://www.nrel.gov/grid/test-case-repository.html.

[99] Leslie J. Managing grid stability in a high IBR network; 2022. https://www.esig.energy/event/webinar-managing-grid-stability-in-a-high-ibr-network/.

[100] Cheng Y, Fan L, Rose J, *et al*. Real-world subsynchronous oscillation events in power grids with high penetrations of inverter-based resources. *IEEE Transactions on Power Systems*. 2022;1–1.

Chapter 2

Real-time testing of smart-WAMS for the monitoring of power oscillation

Lalit Kumar[1], Nand Kishor[2], Merkebu Zenebe Degefa[3], Thuc Dinh Duong[3] and Salvatore D'Arco[3]

This chapter presents the real-time laboratory implementation/testing of smart wide-area monitoring system (smart-WAMS) algorithm. The smart-WAMS is the published approach to monitor power oscillations by the first two authors of this chapter. Smart-WAMS consumes the phasor measurement unit (PMU) signal and gives the accurate real-time information about the critical modes present in the system. However, in this chapter, the smart-WAMS is restructured to track the well-experienced critical mode in real time. The testing is done in national smart grid laboratory (NSGL), SINTEF, Norway under "ERIGrid transnational access" project awarded by European Union-Horizon 2020. The three power system networks are considered for testing and the first one is the "Nordic grid" system, the second is the offline "IEEE-39 bus system," and the last is the "North American SynchroPhasor Initiative (NASPI)" system. The real-time tracking of power oscillations is shown in video results.

2.1 Introduction

The main concern with the rotor angle stability is the power transfer capability in the power system. Therefore, the oscillations associated with rotor angle are generally called as power oscillations [1]. Power system oscillations worsen the reliability, affect the optimal load dispatch, lead to apparatus failure, and may even cause cascaded blackout [2]. Intense power oscillations can propagate throughout the system and are reflected in multiple system variables, some of which are measured by phasor measurement unit (PMU) wherever the PMU is installed. Power oscillations are categorized in Ref. [1]. Two types of power oscillations are very common and severe in power system i.e. "inter-area oscillations" and forced

[1]EED, PEC Chandigarh, India
[2]FCSEE, Østfold University College, Norway
[3]NSGL, SINTEF, Trondheim, Norway

oscillation (FO). Both of these have been extensively studied and addressed in the literature including their origination, properties, consequences, frequency range, countermeasures, and differences in both [1]. To mention again, if the FO frequency is close to the frequency of inter-area mode in the system, then the condition of resonance takes place resulting the significant amplification in the FO [1,2]. Therefore, both of these oscillations are required to be monitored in real time and should be considered seriously in the power system operation and planning.

The necessity of the online "wide-area monitoring" has pushed the authors to estimate the oscillatory dynamics by processing PMUs data. In the line of this necessity, the spectral tools are extensively implemented for mode analysis [3]. However, the spectral contents in the PMU signal are adjacently suffered by heavy ambient noise, multiple events, and closely spaced multiple modes appearing and disappearing with time. Sometimes, low-amplitude–low-frequency modes appear, accompanied with mode-mixing. In such cases, the straight implementation of spectral tools on PMU signal is not proven effective, especially when multiple low-frequency modes are superimposed with heavy ambient noise. Researchers keep on trying to attain the accurate spectral estimation either by developing spectral tool or by incorporating signal processing techniques before spectral estimation [1,3–6].

The research chapter [2] from FinGrid (Finnish power company) presented and analyzed the most significant resonance cases in the Nordic power system by utilizing the PMU signal. The online spectrogram tool in the control center is used for this detection of most significant resonance cases in the past. The online spectrogram tool, which the FinGrid uses, is an old approach, which is just a short-time Fourier transform (STFT) operated on the small segments of the preprocessed signal. This has been theoretically prevailed by many other modified approaches [4,5,8,9].

This chapter presents the initial results with an attempt to make theoretically published "smart wide area monitoring system (smart-WAMS)," practically applicable in industries. The real-time simulator, Opal-RT, is utilized in the laboratory for algorithm testing. The testing is done on three power system networks i.e. "Nordic grid" [10], offline "IEEE-39 bus system" [11], and "North American SynchroPhasor Initiative (NASPI)" system [12]. The smart-WAMS is embedded in the recently developed universal online monitoring MATLAB®-based tool called as Synchromeasurement Application Development Framework (SADF). The procedure and other details involved in the testing are explained and the obtained results are also discussed.

The rest of the chapter is organized as follows: in Section 2.2, the state-of-the-art of the research is presented. Section 2.3 presents the brief discussion about the published smart-WAMS algorithm. Section 2.4 discusses the executed tests and experiments, Section 2.5 shows the results for three power system networks, and lastly, the conclusions are drawn in Section 2.6.

2.2 State-of-the-art

The authors, Zhou *et al.* [4], proposed a self-coherence method to detect FO in the power system. The magnitude-squared coherence (MSC) estimate is incorporated which consumes two signals. The second signal is the delayed signal of the first one. The researchers have also incorporated empirical mode decomposition (EMD) [13] technique before applying spectral tools for well spectral estimation [5,9,14,15]. But sometimes, EMD fails due to its inherent shortcomings such as mode mixings [16,17]. EMD originally belongs to the field of signal processing and the maximum improvements in it are also reported in signal processing field [17–21]. However, the power system researchers also have shown their efforts to make EMD more effective for its application in the power system [5,8,9,22].

After mode decomposition, the decomposed modes may indicate the center frequency in its spectrum, which was not possible without decomposition. Following the same direction, the authors in [22] have proposed a mode decomposition technique called target-EMD (T-EMD), to aid the performance of traditional FFT spectrum in mode estimation. T-EMD is a kind of "two stage mode decomposition (TSMD)" technique, which involves standard EMD and mask-EMD. Similarly, the authors in [23] use EMD to have the trend identification and de-noising of measured power system oscillations. It is stated that the use of nonstationary techniques is more suitable than moving-average approaches in analyzing rapid variations in nonstationary phenomenon characterized by short-lived irregularly occurring events.

Following these motivations, the first two authors of this chapter have published a "TSMD"-based wide-area monitoring approach in Refs. [1,24] to monitor power oscillation. The research stands on the observation that more good decomposition of PMU signal leads to have more good spectral estimation. Two different mode decomposition techniques i.e. CEEMDAN-2014 [18] and VMD [19] are incorporated. Where the reported mode decomposition techniques fail to separate out the closely spaced modes present in the parent signal, the proposed TSMD eliminates the mode-mixing problem in the decomposition process and thus helps in accurate spectral and statistical estimation in decomposed modes. The published wide-area monitoring approach in Refs. [1,24], when implemented here in this chapter in real time, is referred as "smart-WAMS."

2.3   Brief on smart-WAMS

Smart-WAMS is a full package developed to monitor power oscillations, the flow chart of which is shown in Figure 2.1. Smart-WAMS has already been proven effective in Refs. [1,24]. Further to show effectiveness here quickly, let us consider the PMU signal of Hisar bus from the northern regional power grid (NRPG) system of India [25]. Figure 2.2 shows the spectral view of TSMD operated on this considered signal. The power spectral density (PSD) of the said signal is shown in the left most column of Figure 2.2. In the right most column in Figure 2.2, it can be clearly seen that 0.8667 and 0.78 Hz modes which are closely spaced mode have been separated out efficiently by the TSMD and that is why the TSMD is a backbone of smart-WAMS algorithm. The estimation of 1.357 Hz mode in the same signal is tried by other frequency estimation tools and the comparison results are shown in Table 2.1, where it can be clearly seen that TSMD aids the performance of frequency estimation tools to the vast extent.

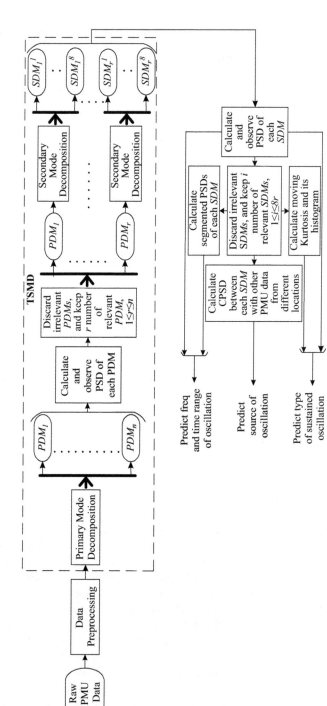

Figure 2.1 Flow chart of proposed monitoring approach, smart-WAMS

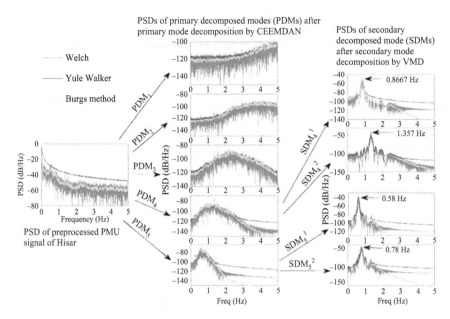

Figure 2.2 Spectrum view of TSMD applied on PMU signal of Hisar

Table 2.1 Comparison in estimation of 1.357 Hz mode by different frequency estimation tools

Approach	Frequency estimation tools	Colourmap for 20 min	Computational time
Without decomposition of parent signal	Segmented-self coherences		3.74 s
	Hilbert transform		1.15 s
	Segmented-PSDs		3.35 s

(Continues)

Table 2.1   (Continued)

Approach	Frequency estimation tools	Colourmap for 20 min	Computational time
Single-stage mode decomposition by CEEMDAN-2014 (PDM$_4$)	Segmented-self coherences	Freq (Hz) vs Time (min) colormap (scale 0–1.5)	19.34 s
	Hilbert transform	Freq (Hz) vs Time (min) colormap (scale −20 to −100)	16.75 s
	Segmented-PSDs	Freq (Hz) vs Time (min) colormap (scale 0 to −100)	18.95 s
Two-stage mode decomposition (SDM$_4{}^2$)	Segmented-self coherences	Freq (Hz) vs Time (min) colormap (scale 0–1)	27.34 s
	Hilbert transform	Freq (Hz) vs Time (min) colormap (scale 0 to −100)	24.75 s
	Segmented-PSDs	Freq (Hz) vs Time (min) colormap (scale −20 to −100)	26.95 s

The proposed approach is for the frequency estimation that also includes the depiction of relative intensity (amplitude) of the mode with respect to normal (or past) intensity as will be seen later in video results. The algorithm does not propose any novel damping estimation approach. The flow chart in Figure 2.1 is only for the offline theoretical depiction of smart-WAMS and not for real-time laboratory testing of smart-WAMS. As can be seen in Figure 2.1, smart-WAMS can provide three information as an output, but in this chapter, smart-WAMS is only tested in the laboratory for the first information i.e. "prediction of frequency and time range of oscillation," which requires access to only one PMU signal. It is well known that the frequency of power oscillations deviates only in limited range across its nominal frequency, which may typically be ±0.1 Hz. Therefore, in order to track the desired oscillation speedily, the algorithm should only process the decomposed modes lying in the desired frequency range. Therefore, the algorithm is written so as to take the desired frequency range as an input to save the computation time.

2.4   Executed tests and experiments

The typical block diagram of the Opal-RT laboratory set-up for wide-area monitoring, protection, and control (WAMPAC) application is already given in Ref. [7] and on other Internet sources. The task of the five-member team (authors) was the "real-time testing of smart-WAMS" in National Smart Grid Laboratory (NSGL), SINTEF, Norway. The task and the team were divided into two subtasks and two subteams, respectively, as shown in the flow chart in Figure 2.3, where subteam 1 consists of authors 3–5 and subteam 2 consists of authors 1 and 2. Although each subteam was mainly engaged in respective assigned subtasks, but both were in regular discussion with each other to resolve the issues and make things work out quickly. The smart-WAMS algorithm consumes the PMU signal in real time to produce the output for visualization in real time. For power oscillation monitoring,

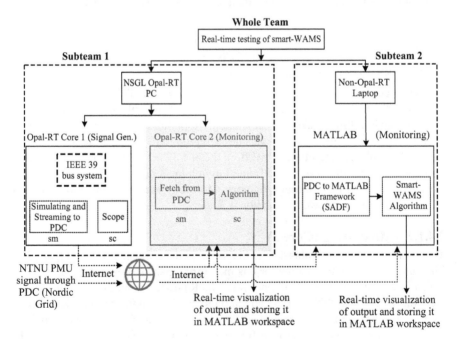

Figure 2.3   Flow chart for real-time testing of smart-WAMS

the test signal should be from high-voltage bus since power oscillation is associated with rotor dynamics. Though the practical offline test cases (PMU signals) for oscillation study can be availed easily either from various Internet sources [26] or by other means, but to get access of PMU signal in real time is not that much easy.

This access requires a huge amount of time in going through the documentary/ agreement work and even then, this access is rarely possible for common researchers. Therefore, the signal from the NTNU campus's PMU installed at the low-voltage (230 V) bus for research purpose is opted for testing, compromising the effective result. The second test case is availed by the real-time simulation of IEEE-39 bus system in a real-time simulator, Opal-RT. Both these test cases were accessed in real time in NSGL during the project visit. The third test case is a PMU signal from NASPI the same as considered in Refs. [1,27]. Unlike first two test cases, the third test case is not accessed in NSGL during the project visit, which will be clarified later.

As depicted in Figure 2.3, the subtask 1 of subteam 1 was to simulate the power system, to stream out the signal in real time, to fetch the streamed out signal in real time and to interface the smart-WAMS algorithm for real-time monitoring. Subtask 1 is executed entirely on the Opal-RT platform. While the subtask 2 of subteam 2 was to fetch the streamed out signal in MATLAB in real time, to rewrite the smart-WAMS algorithm according to the compatibility of SADF framework, and to embed the smart-WAMS algorithm in the SADF framework for real-time monitoring. The subtask 2 is executed completely on Non-Opal-RT platform (MATLAB).

2.4.1   NSGL Opal-RT computer platform: subtask 1

2.4.1.1   Signal generation (Core 1)

For signal generation, IEEE 39 bus system is simulated in ePHASORsim in Core 1 of Opal-RT and the network data is provided in PSS/E file format. The default model given in PSS/E is taken for simulation. To mimic the random load deviation, the ambient noise is given to all loads deviating up to ±5% from the per unit load values. The voltage magnitude signal from the bus-39 is selected and sent to phasor data concentrator (PDC) in the real time through local network. The signal is then streamed out from PDC through Internet. More on data communication for both subtasks will be overviewed in Section 4.3 later.

2.4.1.2   Monitoring (Core 2)

This monitoring part of subtask 1 is the same as the darkened part in Figure 2.3. The outermost RT-Lab window for this monitoring part is shown in Figure 2.4(a), which is executed in Core 2 of Opal-RT. The RT-Lab is configured with PDC to fetch the signal from PDC to RT-Lab via Internet. This configuration is kept inside the master subsystem, "sm_StreamIN," which is able to fetch any signal from PDC and which is streamed out through Internet. Now, the signal fetched by this sub-system is forwarded as an input to console subsystem, "sc_Compute" as depicted in Figure 2.4(a), where the smart-WAMS algorithm is executed in real time

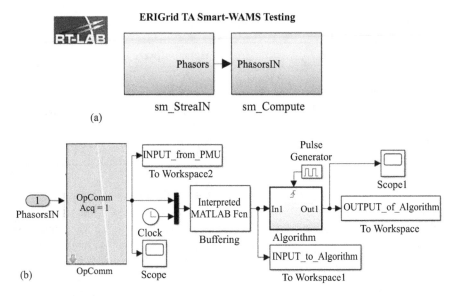

Figure 2.4 *(a) Outermost block diagram in RT-Lab for subtask 1. (b) Inside block diagram of sc_Compute subsystem*

consuming the input signal in real time. The block "sc_Compute is explored in Figure 2.4(b), which consists of following important blocks:

- **OpComm:** To receive the signal from another block
- **Interpreted MATLAB function:** To bypass any MATLAB function from RT-Lab to MATLAB for its execution in MATLAB. Here the buffer function is bypassed.
- **Pulse generator:** To start the algorithm after each received window frame.
- **To workspace:** To store the relevant input/output to MATLAB workspace after the experiment termination.

2.4.2 Non-Opal-RT laptop platform: subtask 2

In this platform, a non-Opal-RT laptop is used to fetch the streamed out signal to MATLAB in real time. The most important tool in this platform is SADF [28]. SADF is a MATLAB coding-based library, which enables receiving of TCP, UDP or TCP/UDP synchro-measurement data. SADF is not only for receiving the signal but it has a broad application in WAMPAC [28]. The downloaded SADF library [28] should be added to the MATLAB path to be able to utilize it. The function "SADF_setting" allows to configure the PMU/PDC connection settings. The function "SADF_run" allows to embed any designed WAMPAC algorithm in it, just like the signal fetching algorithm, "demo_WAMS" is by default embedded in it. To make any WAMPAC algorithm compatible for embedding in "SADF_run" requires the understanding of the function, "demo_WAMS."

Table 2.2   Major hardware and software involved

Major hardware involved	Major software involved
• Computer, and Laptop with 6 GB RAM, 3.2 GHz, Window 7 (64 bit) • Opal-RT (ePHASORSIM) • PMU • PDC	• MATLAB 2016 • PSS/E • Purchased code to fetch the streamed out signal to RT-Lab • SADF library to fetch the signal to MATLAB

In this subtask 2, the algorithm is recoded according to the compatibility of SADF, which is given the filename as "smart_WAMS_embed." This function is embedded into the main function, "SADF_run." Now there are two functions embedded in the main function, "demo_WAMS" and "smart_WAMS_embed." The function "demo_WAMS" is given by default in SADF, which is to plot the fetched signal in real time with its specifications mentioned on the plot. And, the function "smart_WAMS_embed" is to plot the figure for the dominant mode monitoring as will be seen in Section 2.5. The function "SADF_run" is edited so that the embedded functions, "demo_WAMS" and "smart_WAMS_embed," will plot the two figures each of which will be displayed/fitted in the right and left half of the laptop screen, respectively. And, the results are displayed in real time.

2.5   Data communication and storage

The PMU signal is streamed according to IEEE C37.118.2-2011 compliance with a specific host ID (TCP/IP), port, and device ID code. The signal is streamed/received at the sampling frequency of 50 Hz and 60 Hz for test cases 1 and 2 and test case 3, respectively. The availability of the stream was checked by the open source software called PMU connection tester [29]. If desired, the real-time signal and monitoring outputs may be stored in the workspace of MATLAB/RT-Lab in both subtasks. However, storing will limit the monitoring time since the computer will hang after workspace is loaded with heavy data. The function, "smart_WAMS_embed," is written in such a way that whichever power oscillation (frequency) will be most dominant in provided frequency range will only be displayed on the screen (colormap). In overall "real-time testing of smart-WAMS," the major hardware and software used are given in Table 2.2.

2.6   Results

As mentioned before that the monitoring part was performed on two platforms, but later it turned out that for monitoring purpose, the Opal-RT platform (darkened part in Figure 2.3) was complex and ineffective in comparison to the non-Opal-RT (MATLAB) platform. Therefore, the darkened part in Figure 2.3, i.e. Section 4.1.2

can be ignored in the "Real-time testing of smart-WAMS" and thus the monitoring results are only shown from the Non-Opal-RT platform in this section. To interpret here, if the access to PMU signal is available in real time through the Internet, then it can be monitored in real time from anywhere in the world without the need of Opal-RT at all.

2.6.1   PMU signal from Nordic grid: test case 1

In general, the identification of inter-area oscillation is difficult to show in comparison to FO, as it is the ambient characteristic with low-amplitude whereas FO has relatively high amplitude with almost negligible damping and negligible frequency deviation. That is why the spectrogram in Ref. [2] does not show the inter-area mode identification prior to the resonance in their sixth figure. Further to notice from their seventh figure, that when resonance takes place, the deviating frequency stops deviating and remains nearly constant. In Nordic grid, frequency of one inter-area mode deviates in between 0.3 and 0.4 Hz in which the machines of southern Norway and southern Sweden oscillate against the machines of southern Finland [2]. With the same mode of interest for tracking, the low-voltage PMU signal from NTNU-campus's bus is monitored in real time, which belongs to the Nordic grid. The video result for the real-time monitoring of deviating 0.33 Hz inter-area mode in voltage magnitude signal is shown in Ref. [30]. The same mode has been studied several times in literature [31]. The snapshot from the video result is also shown in Figure 2.5. The size of the receiving window should not be too small as mode decomposition works better with large-length dataset. This size is kept to 60 s, which is downsampled from 50 Hz to 5 Hz by the algorithm. The overlapping (sliding) time-frame for each next window is determined by the computation time taken by the previous window. And, the computation time for every

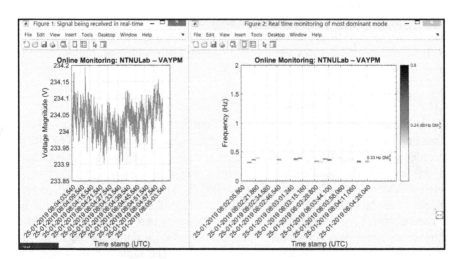

Figure 2.5   Real-time monitoring of low-voltage Nordic grid PMU signal for inter-area mode [30]

window is not fixed, mainly because the primary decomposed modes obtained by CEEMDAN-2014 (in Smart-WAMS) are not fixed every time. In addition, if there is no mode detected in the provided frequency range, the algorithm will skip the computation, consuming nearly no time and will jump to the latest received 60 s window. Therefore, the overlapping time-frame is not constant throughout, which can be noticed from ununiformed gap in the time stamps on colormap shown in video results. In this case of Nordic grid, the input frequency range that is provided to the algorithm is [0.25 0.45] Hz. It can be seen that even in the low-voltage PMU signal, the smart-WAMS is able to track the inter-area mode in an ambient data in real time.

The time stamp on x-axis does not mean that corresponding y-axis result is for that particular time stamp in colormap. The x-axis time stamp is the middle value of the window of 3,000 samples (300 downed samples) i.e. the corresponding y-axis result is for time stamp ±30 s. It is so as to avoid conjunction on the x-axis. DM_i^j is referred as jth secondary decomposed mode (DM) obtained after secondary decomposition operated on the ith primary DM. Readers can follow Refs. [1,24] for more on smart-WAMS. The offline analysis of PMU data for inter-area mode and FO identification has already been carried out in Refs. [1,24].

2.6.2   *Simulated signal from IEEE-39 bus system: test case 2*

The author [32] has studied a 0.9898 Hz inter-area mode with (G2, G3) opposite to (G4, G5, G6, G7) in offline IEEE-39 bus system. The same system is simulated in Opal-RT with the same mode of interest. The algorithm is run on the 39th bus voltage magnitude signal, being received in real time by the algorithm through Internet via TCP/IP and PDC port setting. In a similar manner as in the previous section, the video result is obtained and shown in Ref. [33]. The snapshot from the video is also shown in Figure 2.6. The 3-phase fault is given and cleared at 39th bus at time 14:50:36 in real time. It can be seen that under normal ambient case, the inter-area mode is difficult to observe but as the disturbance occurs, the intensity of the inter-area mode increases and comes down gradually with time. In this case, the input tracking frequency range is kept to [0.85 1.5] Hz.

2.6.3   *Offline PMU signal from NASPI: test case 3*

Due to the absence of FO in the test cases 1 and 2, the tracking of FO was missed during the project visit to NSGL. Although offline identification of FO (4.62 Hz) in recorded PMU signal from NASPI is shown in Refs. [1,27] by the first two authors of this chapter, the need/curiosity of similar real-time tracking of FO through smart-WAMS is felt. Therefore, it is decided to do alteration in SADF so that instead of fetching the streamed out signal from TCP/IP and port, the signal can be fetched to "smart_WAMS_embed" from the workspace in real time. The embedded algorithm, "smart_WAMS_embed," would now produce the real-time results in exactly the same manner as for test cases 1 and 2. The same NASPI's signal is considered as in Refs. [1,27] and the real-time tracking of 4.62 Hz FO is shown in video result in Ref. [34]. The snapshot of result is also shown in Figure 2.7.

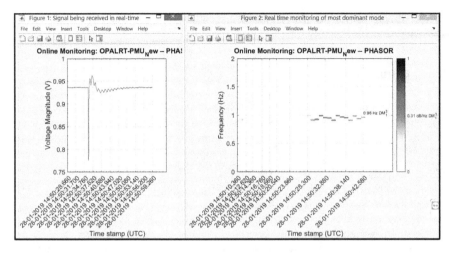

Figure 2.6   Real-time monitoring of Opal-RT PMU signal for inter-area mode (IEEE-39 bus system) [33]

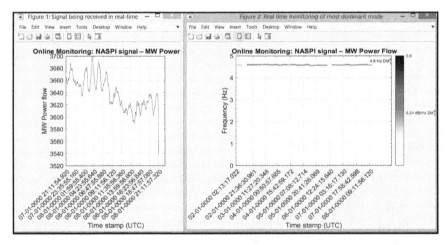

Figure 2.7   Real-time monitoring of NASPI's PMU signal for FO [34]

It is to mention here that the time stamps information of this 10-min PMU signal was kept confidential and was not made public. Unlike, in low-frequency oscillation (LFO) case, the downsample factor cannot be kept high in high-frequency FO case. Therefore, for fixed time-window, the data samples are more for FO tracking in comparison to LFO tracking. As a result, the speed of FO tracking would definitely be slower than LFO tracking.

To avoid the resonance, the usual practice is to keep the frequency of forced oscillation away from the inter-area oscillation [1]. By effective tracking with smart-WAMS, closely spaced inter-area mode and FO can be thoroughly monitored in real time and the necessary precaution can be taken if the risk of resonance is seen.

2.7   Conclusions

Real-time laboratory implementation/testing of the published wide area monitoring approach, smart-WAMS is executed in NSGL, SINTEF, Norway. The real-time simulator, Opal-RT, is utilized in the laboratory for algorithm testing. Smart-WAMS is embedded in MATLAB-based online monitoring tool, SADF, which receives the PMU signal from PDC through Internet via TCP/IP protocol. Smart-WAMS is the first algorithm, which is embedded in SADF since the SADF is made publically available.

The video results for power oscillation tracking in Nordic grid system, offline IEEE-39 bus system, and NASPI system are shown in real time. Smart-WAMS can help in minimizing the risk of instability in the power system by tracking the critical oscillation in real time and thereby taking the corrective action if significant frequency deviation in closely spaced modes is observed.

Acknowledgment

This work has been carried out under the project awarded by European Union-Horizon 2020 (H2020) under the scheme of ERIGrid transnational access. The authors acknowledge the support received from the coauthors associated with NSGL, SINTEF, Norway.

References

[1]   Kumar, L. and Kishor, N. 'Wide area monitoring of sustained oscillations using double-stage mode decomposition', *Int. Trans. Electr. Energy Syst.*, 2018, 28(6), pp. 1–18.

[2]   Seppanen, J., Turunen, J., Nikkila, A.J., and Haarla, L. 'Resonance of forcing oscillations and inter-area modes in the Nordic power system', in: *Proceedings – 2018 IEEE PES Innovative Smart Grid Technologies Conference Europe*, ISGT-Europe 2018, Institute of Electrical and Electronics Engineers Inc., 2018.

[3]   Vanfretti, L., Bengtsson, S., and Gjerde, J.O. 'Preprocessing synchronized phasor measurement data for spectral analysis of electromechanical oscillations in the Nordic grid', *Int. Trans. Electr. Energy Syst.*, 2015, 25(2), pp. 348–558.

[4]   Zhou, N. and Dagle, J. 'Initial results in using a self-coherence method for detecting sustained oscillations', *IEEE Trans. Power Syst.*, 2015, 30(1), pp. 522–530.

[5]   Yang, D., Li, Y., Rehtanz, C., and Yang, D. 'A hybrid method and its applications to analyse the low frequency oscillations in the interconnected power system', *IET Gener. Transm. Distrib.*, 2013, 7(8), pp. 874–884.

[6] Chen, L., Min, Y., and Hu, W. 'An energy-based method for location of power system oscillation source', *IEEE Trans. Power Syst.*, 2013, 28(2), pp. 828–836.

[7] Vanfretti, L., Chenine, M., Almas, M.S., Leelaruji, R., Angquist, L., and Nordstrom, L. 'SmarTS Lab – a laboratory for developing applications for WAMPAC systems', in: *IEEE Power and Energy Society General Meeting*, 2012.

[8] Laila, D.S., Messina, A.R., and Pal, B.C. 'A refined Hilbert-Huang transform with applications to interarea oscillation monitoring', *IEEE Trans. Power Syst.*, 2009, 24(2), pp. 610–620.

[9] Yang, D., Rehtanz, C., Li, Y., and Tang, W. 'A novel method for analyzing dominant oscillation mode based on improved EMD and signal energy algorithm', *Sci. China Technol. Sci.*, 2011, 54(9), pp. 2493–2500.

[10] Chompoobutrgool, Y., Vanfretti, L., Wei Li, and Vanfretti, L. 'Development and implementation of a Nordic grid model for power system small-signal and transient stability studies in a free and open source software', 2012 *IEEE Power Energy Soc. Gen. Meet.*, 2012, pp. 1–8.

[11] Pal, B.C. and Singh, A. Benchmark Systems for Stability Controls Report on the 68-Bus, 16-Machine, 5-Area System, 2013.

[12] Dagle, J. 'North American SynchroPhasor initiative – an update of progress', in: '*Proceedings of the Annual Hawaii International Conference on System Sciences*, Big Island, HI, 5–8 January 2009, pp. 1–5.

[13] Huang, N.E., Shen, Z., Long, S.R., *et al.* 'The empirical mode decomposition and the Hilbert spectrum for nonlinear and non-stationary time series analysis', *Proc. R. Soc. London A*, 1998, 454(1971), pp. 903–995.

[14] Trudnowski, D.J. 'Estimating electromechanical mode shape from synchrophasor measurements', *IEEE Trans. Power Syst.*, 2008, 23(3), pp. 1188–1195.

[15] Sanchez-Gasca, J., D. Trudnowski, E. Barocio, *et al.* 'Identification of electromechanical modes in power systems', in: IEEE Task Force Report, Special Publication TP462, IEEE Power and Energy Society, 2012.

[16] Jingsong, L., Quan, L., and Hang, S. 'The study of the intermittency test algorithm to eliminate mode mixing', in: *Proceedings – 4th International Congress on Image and Signal Processing*, CISP 2011, Shanghai, China, 15–17 October 2011, pp. 2384–2387.

[17] Tang, B., Dong, S., and Song, T. 'Method for eliminating mode mixing of empirical mode decomposition based on the revised blind source separation', *Signal Process.*, 2012, 92(1), pp. 248–258.

[18] Colominas, M.A., Schlotthauer, G., and Torres, M.E. 'Improved complete ensemble EMD: a suitable tool for biomedical signal processing', *Biomed. Signal Process. Control*, 2014, 14(1), pp. 19–29.

[19] Dragomiretskiy, K. and Zosso, D. 'Variational mode decomposition', *IEEE Trans. Signal Process.*, 2014, 62(3), pp. 531–544.

[20] Gilles, J. 'Empirical wavelet transform', *IEEE Trans. Signal Process.*, 2013, 61(16), pp. 3999–4010.

[21] Ryan Deering, J.F. Kaiser, D. R., and Kaiser, J.F. 'The use of a masking signal to improve empirical mode decomposition', in: *IEEE International Conference on Acoustics, Speech, and Signal*, Philadelphia, PA, USA, 23–23 March 2005, IEEE, pp. 485–488.

[22] Prince, A., Senroy, N., and Balasubramanian, R. 'Targeted approach to apply masking signal-based empirical mode decomposition for mode identification from dynamic power system wide area measurement signal data', *IET Gener. Transm. Distrib.*, 2011, 5(10), pp. 1025–1032.

[23] Messina, A.R., Vittal, V., Heydt, G.T., and Browne, T.J. 'Nonstationary approaches to trend identification and denoising of measured power system oscillations', *IEEE Trans. Power Syst.*, 2009, 24(4), pp. 1798–1807.

[24] Kumar, L. and Kishor, N. 'Determination of mode shapes in PMU signals using two-stage mode decomposition and spectral analysis', *IET Gener. Transm. Distrib.*, 2017, 11, pp. 4422–4429.

[25] 'NRLDC – Reliability, Security and Economy in Power Transmission', https://nrldc.in/, accessed December 2015.

[26] Maslennikov, S., Wang, B., Zhang, Q., *et al.* 'A test cases library for methods locating the sources of sustained oscillations', in: *IEEE Power and Energy Society General Meeting*, IEEE Computer Society, 2016.

[27] Kumar, L. and Kishor, N. 'Frequency monitoring of forced oscillation in PMU's data from NASPI', in: *Proceedings of the 18th Mediterranean Electrotechnical Conference*, Limassol, Cyprus, 18–20 April 2016, IEEE, pp. 18–20.

[28] Naglic, M., Popov, M., van der Meijden, M.A.M.M., and Terzija, V. 'Synchro-measurement application development framework: an IEEE standard C37.118.2-2011 supported MATLAB library', *IEEE Trans. Instrum. Meas.*, 2018, 67(8), pp. 1804–1814.

[29] 'PMU Connection Tester', https://archive.codeplex.com/?p=pmuconnection-tester, accessed February 2019.

[30] 'Online Inter Area Mode Monitoring in Nordic Grid System', https://drive.google.com/file/d/1o76Tkv9lKLRmu7h1mU4KS9X4y2G5FKZQ/view?usp=sharing, accessed January 2019.

[31] Vanfretti, L., Bengtsson, S., Peric, V.S., and Gjerde, J.O. 'Spectral estimation of low-frequency oscillations in the Nordic grid using ambient synchrophasor data under the presence of forced oscillations', in: 2013 *IEEE Grenoble Conf. PowerTech*, POWERTECH 2013, 2013, pp. 1–6.

[32] Yu, Y., Grijalva, S., Thomas, J.J., Xiong, L., Ju, P., and Min, Y. 'Oscillation energy analysis of inter-area low-frequency oscillations in power systems', *IEEE Trans. Power Syst.*, 2016, 31(2), pp. 1195–1203.

[33] 'Online Inter Area Mode Monitoring in IEEE 39 Bus System', https://drive.google.com/file/d/1fIU8N-ju1XblcA7GVKEsJzP9wuXE8jcC/view?usp=sharing, accessed January 2019.

[34] 'Online FO Monitoring in NASPI System', https://drive.google.com/file/d/1rEK4rFP_tYvMD9RCWIv41FhrCfO4tLu1/view, accessed January 2020.

Chapter 3

Wide-area control design in different aspects of oscillations

Priyatosh Mahish[1] and Sukumar Mishra[2]

3.1 Introduction

With gradual increase in the interconnection of power transmission systems, monitoring and control of the network raises new challenges. It is now approved by the power engineers that the local control techniques cannot always deliver sufficient damping [1]. To overcome such issues, synchrophasor data-based control (SDC) methods have proven their potential in several power system applications [2]. In these methods, using global positioning system (GPS), phasor measurement units (PMUs) generate time-synchronized measurements of their respective substations. The synchronized/synchrophasor data are aggregated to the phasor data concentrator (PDC) through communication channels. Further, PDC uses these data to obtain wide-area control (WAC) logics in a control center [1].

3.1.1 WAC applications in renewable integrated power system

- Renewable energy resources are acknowledged to be only option throughout the globe for sustainable growth of power system. To reduce the environmental pollution and global warming, fossil fuels will be obsolete in near future. As a result, synchronous generator (SG)-based power plants are replaced with the converter-based generation stations. Thereby, system inertia is reduced significantly leading to stability issues. Under this circumstance, several countries are modifying the grid codes to enhance adaptability of power system with the behavior of renewable resources. As an example, large-scale wind farms (WFs) (few hundreds of megawatts) are mandated to participate in ancillary services. Hence, available power reserve in these WFs can be employed in improving primary frequency regulation (PFR) of the transmission networks [3]. To implement such idea, fast coordination among droop control of WFs

[1]School of Computing and Electrical Engineering, Indian Institute of Technology Mandi, India
[2]Department of Electrical Engineering, Indian Institute of Technology Delhi, India

and SGs is necessary, which can be possible with the synchrophasor technology.

- Fixed series compensation (FSC) and/or thyristor controller series compensation (TCSC) are well-known tools to enhance the loadability of AC transmission lines, which further improves voltage stability and angle stability of the network. However, for such series-compensated systems, sub-synchronous oscillation (SSO) can be dangerous [4,5]. Many researchers attempted to highlight positive and negative impact of large-scale WFs on SSO. Several literatures illustrated impedance network model (INM) of WF-connected power system for mitigating SSR [6–8]. Further, importance of PMU data in analyzing SSR with the INM tool is suggested [9–11].

- To protect the switching devices in power converters, current limiter is introduced in the control process. As a result, deteriorating short circuit current capacity of the converter-based WFs makes serious problem against transient stability and voltage stability of the power network [12,13]. Further, unity power factor operation of these converters raises scarcity of useful reactive power injection to the grid [13]. To handle such issue, large-scale WF-integrated high-voltage substations can coordinate with each other to improve voltage oscillation damping of power system transmission networks. The WAC is well acknowledged for this operation [14,15].

3.1.2   WAC challenges

- The WAC may associate significant delay in communication which affects the stability of the controller [16]. IEEE standard C37.118.2 explains, the time delay introduced in WAC is categorized into different parts such as [17,18]: processing delay in PDC and PMUs, delay in communication medium, and delay in application. Further, communication delay is divided into three parts [19,20]: data queuing delay, propagation delay, and transmission delay. The range of communication delay can be 10 ms to even few seconds [21]. The main factors identified for such wide variations in communication latency are different types of communication medium, transmission protocol, routing of signal transmission, and sudden increase in communication loads [19,22]. Such delay variations may cause disordering in data sequence, which raises a significant challenge for WAC applications [23].

- Message collision or node failure causes packet drop event in WAC [24,25]. Such type of phenomena can be disaster if exists for few seconds.

3.2   WAC methodology

Figure 3.1 shows centralized WAC model, which incorporates two components of WAC. Each component is assigned with a PMU. The time-synchronized data from the PMUs are sent to PDC for obtaining $y1$ and $y2$. The PDC sends these data to WAC. y_1^c and y_2^c are control output vector matrices of the WAC component-1 and

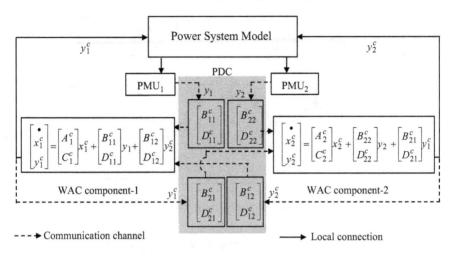

Figure 3.1   Centralized WAC model

component-2, respectively. Using component-1 and component-2, centralized WAC in the control center is formulated as:

$$\begin{bmatrix} \dot{x}_1^c(t) \\ \dot{x}_2^c(t) \end{bmatrix} = \begin{bmatrix} A_1^c + MB_{12}^c C_1^c D_{21}^c & MB_{12}^c C_2^c \\ MB_{21}^c C_2^c & A_2^c + MB_{21}^c C_2^c D_{12}^c \end{bmatrix} \begin{bmatrix} x_1^c(t) \\ x_2^c(t) \end{bmatrix}$$

$$+ \begin{bmatrix} B_{11}^c + MB_{12}^c D_{11}^c D_{21}^c & MB_{12}^c D_{22}^c \\ MB_{21}^c D_{11}^c & B_{22}^c + MB_{21}^c D_{22}^c D_{12}^c \end{bmatrix} \begin{bmatrix} y_1(t) \\ y_2(t) \end{bmatrix} \quad (3.1)$$

$$\begin{bmatrix} \dot{y}_1^c(t) \\ \dot{y}_2^c(t) \end{bmatrix} = \begin{bmatrix} MC_1^c & MD_{12}^c C_2^c \\ MD_{21}^c C_1^c & MC_2^c \end{bmatrix} \begin{bmatrix} x_1^c(t) \\ x_2^c(t) \end{bmatrix}$$

$$+ \begin{bmatrix} MD_{11}^c & MD_{12}^c D_{22}^c \\ MD_{21}^c D_{11}^c & MD_{22}^c \end{bmatrix} \begin{bmatrix} y_1(t) \\ y_2(t) \end{bmatrix} \quad (3.2)$$

where, $M = (1 - D_{12}^c D_{21}^c)^{-1}$ superscript c signifies controller information. x_1^c and x_2^c are the state vectors of WAC component-1 and component-2, respectively. Equations (3.1) and (3.2) can be re-expressed as

$$\begin{aligned} \dot{x}^c &= A^c x^c(t) + B^c y(t) \\ y^c(t) &= C^c x^c(t) + D^c y(t) \end{aligned} \quad (3.3)$$

where the size of WAC state matrix $x^c(t)$ is $(hw \times 1)$, WAC input matrix $y(t)$ is $(w \times 1)$, and WAC output matrix $y^c(t)$ is $(w \times 1)$. $A^c \in \mathbb{R}^{wr \times rh}$, $B^c \in \mathbb{R}^{wh \times q}$, $C^c \in \mathbb{R}^{w \times hw}$, and $D^c \in \mathbb{R}^{w \times w}$. h is the WAC order, and w is the number of WAC components.

3.2.1   Resilient WAC design

Using (3.1), B_{12}^c, B_{21}^c, D_{12}^c, D_{21}^c modifies the eigen values of the wide-area state transition matrix A^c to optimal enhancement of the WAC stability. Figure 3.2

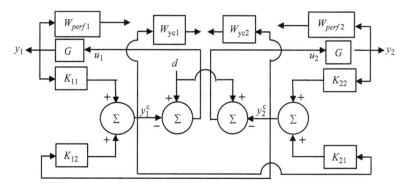

Figure 3.2   Signal flow graph of WAC stability model

depicts the signal flow graph of WAC stability model. The WAC component-related transfer functions are expressed as [26]

$$K_{11}(s) = D_{11}^c + B_{11}^c (Is - A_1^c)^{-1} C_1^c \tag{3.4}$$

$$K_{22}(s) = D_{22}^c + B_{22}^c (Is - A_2^c)^{-1} C_2^c \tag{3.5}$$

$$K_{12}(s) = D_{12}^c + B_{12}^c (Is - A_1^c)^{-1} C_1^c \tag{3.6}$$

$$K_{21}(s) = D_{21}^c + B_{21}^c (Is - A_1^c)^{-1} C_1^c \tag{3.7}$$

G represents reduced-order power system transmission network. The full-order model is reduced with truncated balanced technique [27]. d is a disturbance in the network. The relationship among $d(s)$, $y_1(s)$, and $y_2(s)$ in Figure 3.2 is expressed as

$$(G^{-1} + K_{11}M_1)y_1 = d - (K_{12}K_{22}M_1)y_2 \tag{3.8}$$

$$(K_{11}K_{21}M_1)y_1 = d - (G^{-1} + K_{22}M_1)y_2 \tag{3.9}$$

where $M_1 = (1 - K_{12}K_{21})^{-1}$.

Relations (3.8) and (3.9) are simplified and re-expressed as

$$\begin{bmatrix} y_1(s) \\ y_2(s) \end{bmatrix} = \begin{bmatrix} f_{dy1}(s) \\ f_{dy2}(s) \end{bmatrix} [d(s)] \tag{3.10}$$

Further, the relations between $d(s)$, $y_1^c(s)$, and $y_2^c(s)$ are obtained from Figure 3.2 and are expressed as

$$\begin{bmatrix} y_1^c(s) \\ y_2^c(s) \end{bmatrix} = N_1 \begin{bmatrix} K_{11}f_{dy1} + K_{12}K_{22}f_{dy2} \\ K_{22}f_{dy2} + K_{21}K_{11}f_{dy1} \end{bmatrix} [d(s)] \tag{3.11}$$

By H_∞ optimization process of (3.10) and (3.11), optimal K12 is solved using the objectives:

$$\min_{K_{12} \in S} \left\| \frac{W_{perf}f_{dy1}}{W_{yc1}(K_{11}f_{dy1} + K_{12}K_{22}f_{dy2})M_1} \right\|_\infty < 1 \tag{3.12}$$

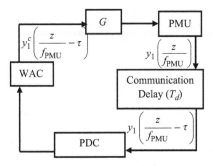

Figure 3.3 WATDS model

$$\min_{K_{21} \in S} \left\| \begin{array}{c} W_{perf2} f_{dy2} \\ W_{yc2}(K_{22} f_{dy2} + K_{11} K_{21} f_{dy1}) M_1 \end{array} \right\|_\infty < 1 \qquad (3.13)$$

S is a set of stabilized controllers. $W_{perf1} = 100/(s + 100)$, $W_{perf2} = 100/(s + 100)$, $W_{yc1} = 0.423s(s + 1.05)$, and $W_{yc2} = 0.423s(s + 1.05)$ [28]. The solutions of (3.12) and (3.13) provide optimal transfer functions of K_{12} and K_{21}, respectively.

3.2.2 Wide-area-time-delayed system model

Let us consider that PDC receives data from PMU_1 with communication latency τ. The wide-area-time-delayed system (WATDS) model is shown in Figure 3.3. f_{PMU} is PMU reporting frames per second (FPS). (z/f_{PMU}) is the time when PMU measures the zth data. Equation (3.14) is used to express the model of WATDS:

$$\begin{aligned} x_c\left(\frac{z+1}{f_{PMU}}\right) &= A^c x_c\left(\frac{z}{f_{PMU}}\right) + B^c y_1\left(\frac{z}{f_{PMU}} - \tau\right) \\ y_1^c\left(\frac{z}{f_{PMU}} - \tau\right) &= C^c x_c\left(\frac{z}{f_{PMU}}\right) + D^c y_1\left(\frac{z}{f_{PMU}} - \tau\right) \end{aligned} \qquad (3.14)$$

3.3 Wide-area-predictive-control approach

The set of model equations of the WAPC is [29]

$$\begin{aligned} x_c\left(\frac{z+1}{f_{PMU}}\right) &= A^c x_c\left(\frac{z}{f_{PMU}}\right) + B^c y_1\left(\frac{z}{f_{PMU}} \Big| \frac{z-m}{f_{PMU}}\right) \\ \widehat{y}_1^c\left(\frac{z}{f_{PMU}}\right) &= C^c x_c\left(\frac{z}{f_{PMU}}\right) + D^c y_1\left(\frac{z}{f_{PMU}} \Big| \frac{z-m}{f_{PMU}}\right) \end{aligned} \qquad (3.15)$$

Figure 3.4 depicts WATDS model with the predictive controller. $y_1\left(\frac{z}{f_{PMU}} \Big| \frac{z-m}{f_{PMU}}\right)$ signifies the $(z-m)$th instant predicted output of zth data associated with unknown communication latency τ. $\widehat{y}_1^c\left(\frac{z}{f_{PMU}}\right)$ is the output of the predictive controller-associated WAC.

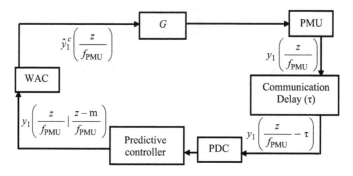

Figure 3.4   WATDS model with predictive controller

3.3.1 Reordering of data frames and data loss compensation at the PDC

Step 1. At PDC, update $z = z + 1$ that acknowledge a data frame from the PMU at zth instant.

Step 2. Identify the dataset from individual PMUs, and order each set with the $(k-1)$ old frames and present frame with associated delays, as $[X_z] \rightarrow \{y_1(z), y_1(z-1), \ldots, y_1(z-k+1)\}$. In the case of single/multiple data drop within the interval of $((-T_{PDC} + z/f_{PMU}), (-1/f_{PMU} + z/f_{PMU}))$, the predicted data are used in order. T_{PDC} is the maximum waiting time of PDC till the data frame is arrived instantaneously or with communication delay. If the associated delay of the data is more than T_{PDC}, then PDC decides to drop that data.

3.3.2 Computational method at the predictive controller

Step 1. Fix the size of $[X_z]$ is $k \times 1$. The condition $k > p > (f_{PMU}T_{PDC})$ must be satisfied at every instant of prediction, where p is the prediction horizon.

Step 2. Obtain η by an iterative method with the following relations:

$$\eta + \eta^2 + \ldots + \eta^n = 1; 0 < \eta < 1 \tag{3.16}$$

Step 3. At zth instant, perform the prediction till the prediction horizon p:

$$y_1((z+1)|z) = \eta y_1(z) + \eta^2 y_1(z-1) + \ldots + \eta^n y_1(z-k+1) \tag{3.17}$$

$$\begin{aligned} y_1((z+2)|z) &= \eta y_1((z+1)|z) + \eta^2 y_1(z) + \ldots + \eta^n y_1(z-k+2) \\ &= 2(\tau^2 y_1(z) + \ldots + \tau^n y_1(z-n+2)) + (2^1-1)\tau^{n+1} y_1(z-k+1) \end{aligned} \tag{3.18}$$

$$\begin{aligned} y_1((z+3)|z) &= \eta y_1((z+2)|z) + \eta^2 y_1((z+1)|z) + \eta^3 y_1(z) \\ &\quad + \ldots + \eta^n y_1(z-k+3) = 4(\tau^3 y_1(z) + \ldots + \eta^n y_1(z-k+3)) \\ &\quad + (2^2-1)\eta^{n+1} y_1(z-k+2) \\ &\quad + (2^1-1)\eta^{n+2} y_1(z-k+1) \end{aligned} \tag{3.19}$$

Similarly,

$$y_1((z+p)|z) = \eta y_1((z+p-1)|z) + \ldots + \eta^{p-1} y_1((z+1)|z) + \eta^p y_1(z)$$
$$+ \ldots + \eta^n y_1(z-k+p) \quad = 2^{p-1}(\eta^p y_1(z) + \ldots + \eta^n y_1(z-k+p))$$
$$+ (2^{p-1}-1)\eta^{n+1} y_1(z-k+p-1)$$
$$+ \ldots + (2^1-1)\eta^{n+p-1} y_1(z-k+1) \tag{3.20}$$

3.3.3 Control horizon setting at WAPC

The control horizon setting is necessary before transmitting predicted data to WAC. For every new z, optimum control horizon $\alpha_{\text{opt}} \leq p$ is required to satisfy. The predictive error increases with the larger α_{opt}. Further, if $p < \alpha_{\text{opt}} < (f_{\text{PMU}} T_{\text{PDC}})$, then the WAPC fails to provide predicted data. In such situation, compensation of delay or/and data losses is not possible. Therefore, optimal control horizon setting is necessary. This optimal setting in predictive controller is obtained as follows.

    Step 1. Initialize $\alpha = 1$.
    Step 2. If $\alpha \geq (T_{\text{wait}}/T)$, then set $\alpha_{\text{opt}} = \alpha$. Else $\alpha = \alpha + 1$ and go to next step.
    Step 3. If $\alpha \leq p$, then switch to step 2. Else WAPC is unable to compensate delay and data losses.

3.3.4 Case study

Two TCSC-connected New England 39-bus system is considered, as shown in Figure 3.5. The TCSC control model is depicted in Figure 3.6 and the respective parameters are provided in Table 3.1. The TCSC-1 and TCSC-2 are engaged in WAC as earlier discussed in Section 3.2. The effect of communication delays is verified through frequency domain analysis and time domain analysis. Figures 3.7 and 3.8 show bode plots of TCSC-1 and TCSC-2, respectively, engaged in WAC with constant communication latencies. Further, Tables 3.2 and 3.3 illustrate bode plot analysis of WAC for TCSC-1 and TCSC-2, respectively. The magnitude and phase of WAC with no delay and 100 ms delay are about the same in the range of low-frequency oscillation (0.1–2 Hz), as shown in Figures 3.7 and 3.8. It signifies that the effect of delay on performance of WAC is insignificant up to 100 ms. WAC performance becomes marginally stable at the gain cross over frequency (ω_{gc}). Up to 100 ms delay, ω_{gc} lies above the low-frequency oscillation range, as evident from Tables 3.2 and 3.3. Thereby, up to 100 ms, WAC is stable in the frequency range 0.1–2 Hz. For more than 100 ms delay, WAC performance deteriorates. In such situation, ω_{gc} becomes within the low-frequency oscillation range. The results in Tables 3.2 and 3.3 demonstrate that phase margin (PM) becomes lower with higher delays. Thereby, WAC stability margin reduces with larger delays. For modal oscillation frequency higher than the ω_{gc}, the relative magnitude with respect to 0 dB gain becomes negative, as evident from the bode diagrams. However, the relative phase with respect to $-180°$ reference remains positive. Therefore, WAC behaves as unstable for the frequencies above ω_{gc}. This study

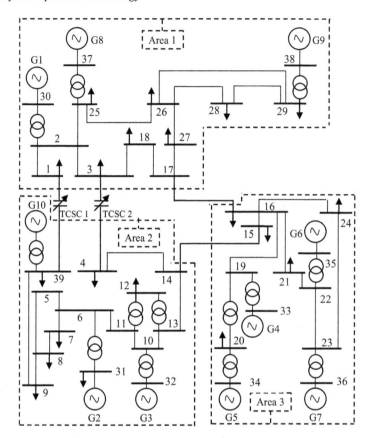

Figure 3.5   New England 39 bus test system

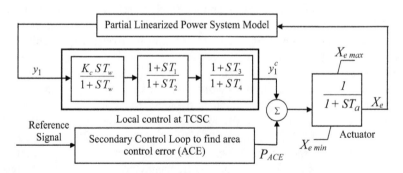

Figure 3.6   TCSC control model

indicates that during low-frequency oscillations, WAC moves toward the instability for increasing communication delays. Figures 3.9 and 3.10 depict oscillations at TCSC-1 and TCSC for constant latencies-associated WAC, respectively. The performance of WAC at both tie-lines, i.e. lines 39-1 and 3-4, are not significantly

Table 3.1   TCSC control parameters

TCSC	T_1	T_2	T_3	T_4	T_w	T_a	K_c
1	0.52	0.1	0.09	0.568	5	0.01	1
2	0.52	0.1	0.09	0.574	5	0.01	1

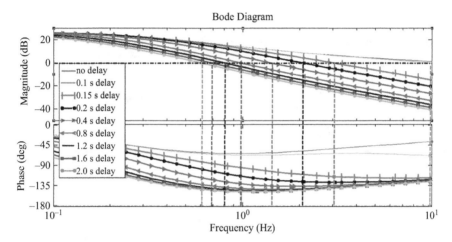

Figure 3.7   Bode plots of TCSC-1 engaged in WAC with constant communication latencies

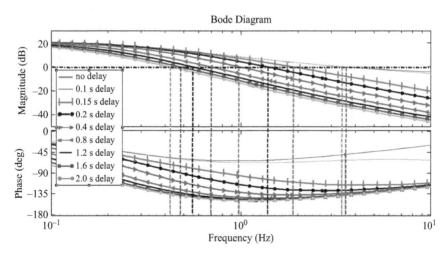

Figure 3.8   Bode plots of TCSC-2 engaged in WAC with constant communication latencies

Table 3.2   Bode plot analysis of WAC for TCSC-1

T_d (ms)	ω_{gc} (Hz)	PM (degree)	T_d (ms)	ω_{gc} (Hz)	PM (degree)
0	14.7	151	800	0.988	38.1
100	9.77	112	1,200	0.799	36.4
200	2.07	53.3	1,600	0.686	36
400	1.42	43.9	2,000	0.609	36.2

Table 3.3   Bode plot analysis of WAC for TCSC-2

T_d (ms)	ω_{gc} (Hz)	PM (degree)	T_d (ms)	ω_{gc} (Hz)	PM (degree)
0	3.53	130	800	0.69	44.4
100	3.4	118	1,200	0.559	43.8
200	1.37	58.1	1,600	0.479	43.5
400	0.979	48.7	2,000	0.424	43

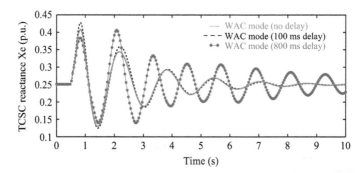

Figure 3.9   Oscillations at TCSC-1 for constant latencies-associated WAC

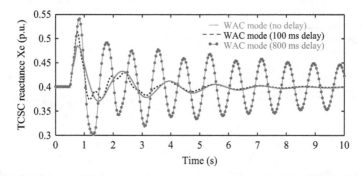

Figure 3.10   Oscillations at TCSC-2 for constant latencies-associated WAC

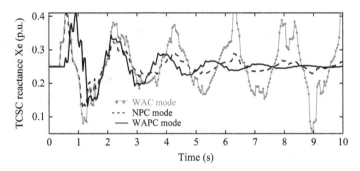

*Figure 3.11   Oscillations at TCSC-1 for different modes of control with random
latencies up to 800 ms and 16% data loss*

affected with 100 ms communication delay, as also explained in frequency domain
analysis. However, associated with 800 ms delay, WAC performance deteriorates
significantly for both the TCSCs.

Figures 3.10 and 3.11 show oscillations at TCSC-1 and TCSC-2, respectively,
for different modes of control with random latencies up to 800 ms and 16% data
loss. The delay distribution is within the range of 100–800 ms with standard
deviation (σ) = 225 ms and average delay (μ) = 442 ms and 16% data packet
drop. The network predictive control (NPC) performs better than the WAC mode
[30]. However, the NPC takes longer time to damp the oscillations, as compared to
the WAPC mode.

3.4   Sub-synchronous oscillation damping in transmission network using synchrophasor technology

This section provides challenges of SSO in WF-integrated transmission network.
Further, the usage of synchrophasor technology in SSO damping for such networks
is elaborated. Several real events on SSO in the transmission power networks are
reported. A 15-MW wind power is purposefully dispatched through a 60% series-
compensated transmission line after an event of line-switching in the area of
Buffalo Ridge at Minnesota [31]. The event results oscillations within a frequency
range of 9–13 Hz. On December 25, 2012, SSO of 6–8 Hz was observed in several
WF-connected transmission lines in North China [32].

3.4.1   Challenges of SSO in WF-integrated transmission networks

SSO raises several issues [33, 34]: (i) electro-magnetic (EM) mode of oscillations is
induced due to induction generator effect (IGE). The frequency of such oscillatory
mode is the compliment of fundamental frequency of system. In another way, it is a
relative difference between electrical resonant frequency of series-compensated

transmission network and fundamental frequency of the system. In a situation where wind generator rotor speed becomes closer to the SSO frequency, the dominance IGE increases. (ii) Mechanical resonance mode can be induced through torsional interaction (TI) at generator–turbine shaft. TI mode is more vulnerable when the network complimentary frequency comes closer to the oscillation frequency of this mode. With increasing compensation, the resonant frequency of electrical transmission network becomes higher resulting more vulnerable TI modes. (iii) An interaction among the electrical resonance and mechanical resonance modes induces resonating torque at the wind generator–turbine shaft, which is called torque amplification (TA) [33]. Sub-synchronous control interaction (SSCI) is another issue for converter-based WFs and series-compensated power systems [35,36]. Due to variations in converter control mechanism, the SSCI mode frequencies may vary in wide range [34]. In 2009, a real event of SSCI is reported at the Electric Reliability Council of Texas (ERCOT) [37]. The event occurred between a 50% series-compensated transmission line and multiple DFIGs.

3.4.2   Application of synchrophasor technology for SSO damping in WF-integrated transmission networks

The impact of real SSO events in the transmission system is challenging to analyze for massive interconnection among the lines and WFs [32]. It is observed that the same SSO mode can change its behavior during transmitting to the WFs connecting to the grid at different buses [38]. To overcome this challenging issue, the impedance network model (INM) is acknowledged as an efficient tool [39,40]. The online INM using synchrophasor technology can be quite effective in implementing INM in online for SSO damping [41]. Coordination among WF-connecting high-voltage substations and other buses in transmission network is possible with PMU data to analyze time-varying nature of the SSO modes in different areas of a grid [42,43]. Figure 3.12 shows the INM model of WF-integrated series-compensated power system. WF_1, WF_2, and WF_3 are connected with a grid through series-compensated transmission lines: line 1, line 2, and line 3 with their impedance Z_{L1}, Z_{L2}, and Z_{L3}, respectively. The line 1 is

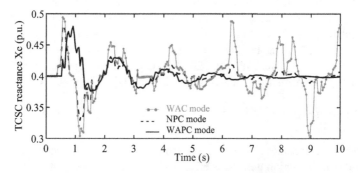

Figure 3.12   Oscillations at TCSC-2 for different modes of control with random latencies up to 800 ms and 16% data loss

compensated with TCSC. Fixed series capacitors (FSCs) are used to compensate lines 2 and 3. I_{L1}, I_{L2}, and I_{L3} are the currents from remote terminals (W_1, W_2, W_3) to the substations/buses (ith, jth, and kth PCC) in transmission network of the grid. The reference impedance of the grid looking from these substations are Z_{eqi}^{rn}, Z_{eqj}^{rn}, and Z_{eqk}^{rn}, respectively. I_{eqi}, I_{eqj}, and I_{eqk} are collective currents from the ith, jth, and kth buses, respectively, to other substations at the transmission network. V_i, V_j, V_k and δ_i, δ_j, δ_k are the voltage magnitudes and angles of the respective WF-connected high-voltage substations. These data are obtained from PMUs. P_{WF1}, P_{WF2}, and P_{WF3} are the grid-connected WF active power injections to the grid. *Egrid* is the voltage at infinite bus, i.e. 1 pu.

By Kirchhoff's voltage law (KVL) at ith and jth WF-connected high-voltage substation,

$$\Delta \vec{Z}_{eqj} = \frac{\vec{y}_{ji} + \vec{I}_{eqi}\Delta \vec{Z}_{eqi}}{\vec{I}_{eqj}} \tag{3.21}$$

where $\vec{y}_{ji} = \vec{V}_i - \vec{V}_j - \vec{I}_{eqi}\vec{Z}_{eqi}^{rn} + \vec{I}_{eqj}\vec{Z}_{eqj}^{rn}$; $\Delta \vec{Z}_{eqj} = \vec{Z}_{eqj}^{rn} - \vec{Z}_{eqj}$; $\Delta \vec{Z}_{eqi} = \vec{Z}_{eqi}^{rn} - \vec{Z}_{eqi}$
Similarly for kth PCC:

$$\Delta \vec{Z}_{eqk} = \frac{\vec{y}_{ki} + \vec{I}_{eqi}\Delta \vec{Z}_{eqi}}{\vec{I}_{eqk}} \tag{3.22}$$

where $\vec{y}_{ki} = \vec{V}_i - \vec{V}_k - \vec{I}_{eqi}\vec{Z}_{eqi}^{rn} + \vec{I}_{eqk}\vec{Z}_{eqk}^{rn}$; $\Delta \vec{Z}_{eqk} = \vec{Z}_{eqk}^{rn} - \vec{Z}_{eqk}$

Using synchrophasor technology, the voltage and current information at the PCCs are measured by the respective PMUs, which are aggregated at the PDC to obtain (3.21) and (3.22). Obtaining such equations for all other PCCs, the PDC forms

$$\Delta \vec{Z}_{eqj} + \dots + \Delta \vec{Z}_{eqk} = \left(\frac{\vec{y}_{ji}}{\vec{I}_{eqj}} + \dots + \frac{\vec{y}_{ki}}{\vec{I}_{eqk}} \right) + \Delta \vec{Z}_{eqi}\vec{I}_{eqi}\left(\frac{1}{\vec{I}_{eqj}} + \dots + \frac{1}{\vec{I}_{eqk}} \right) \tag{3.23}$$

An SDC is required to minimize (4.3) for improving SSO damping at ith PCC. Similarly, SDC minimizes (3.24) and (3.25) for improving SSO damping at jth and kth PCC, respectively:

$$\Delta \vec{Z}_{eqi} + \dots + \Delta \vec{Z}_{eqk} = \left(\frac{\vec{y}_{ij}}{\vec{I}_{eqi}} + \dots + \frac{\vec{y}_{kj}}{\vec{I}_{eqk}} \right) + \Delta \vec{Z}_{eqj}\vec{I}_{eqj}\left(\frac{1}{\vec{I}_{eqi}} + \dots + \frac{1}{\vec{I}_{eqk}} \right) \tag{3.24}$$

$$\Delta \vec{Z}_{eqi} + \dots + \Delta \vec{Z}_{eqj} = \left(\frac{\vec{y}_{ik}}{\vec{I}_{eqi}} + \dots + \frac{\vec{y}_{jk}}{\vec{I}_{eqj}} \right) + \Delta \vec{Z}_{eqk}\vec{I}_{eqk}\left(\frac{1}{\vec{I}_{eqi}} + \dots + \frac{1}{\vec{I}_{eqj}} \right) \tag{3.25}$$

The SDC logic for SSO damping at ith PCC is obtained from (3.23) as

$$|\Delta \vec{Z}_{eqj} + \dots + \Delta \vec{Z}_{eqk}| \leq \left(\frac{|\vec{y}_{ji}|}{|\vec{I}_{eqj}|} + \dots + \frac{|\vec{y}_{ki}|}{|\vec{I}_{eqk}|} \right)$$

$$+ |\Delta \vec{Z}_{eqi}||\vec{I}_{eqi}| \left(\frac{1}{|\vec{I}_{eqj}|} + \dots + \frac{1}{|\vec{I}_{eqk}|} \right) \tag{3.26}$$

In (3.26), $\Delta \vec{Z}_{eqi}$ minimization is performed with local SSR damping controller (SSRDC) K_{11} at ith PCC. The SDC provides K_{12} which is used for the minimization of $y_{eqi} = \left(\frac{|\vec{y}_{ji}|}{|\vec{I}_{eqj}|} + \dots + \frac{|\vec{y}_{ki}|}{|\vec{I}_{eqk}|} \right)$.

Figure 3.13 shows robust SDC model for SSR damping at WF-connected power systems [44]. $y_1(s)$, $d(s)$, and $y_{eq1}(s)$ are related as

$$y_1(s) = \frac{G}{1 + K_{11}G} d(s) + \frac{K_{12}G}{1 + K_{11}G} y_{eqi}(s) \tag{3.27}$$

$$y_1^c(s) = \frac{K_{11}G}{1 + K_{11}G} d(s) + \frac{K_{12}}{1 + K_{11}G} y_{eqi}(s) \tag{3.28}$$

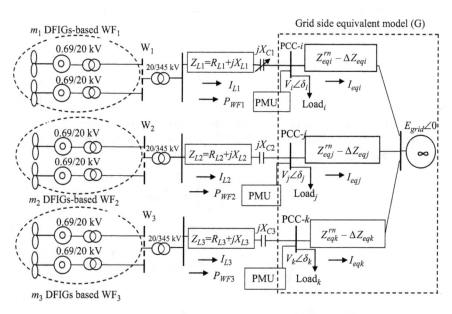

Figure 3.13   INM model of WF-integrated series-compensated power system

Optimization of K_{11} is performed with the objective functions (3.29). Similarly, optimal K_{12} is found from (3.30):

$$\min_{K_{11}\in S} \left\| \frac{\dfrac{W_{perf}G}{1+K_{11}G}}{\dfrac{W_{yc}GK_{11}}{1+GK_{11}}} \right\|_{\infty} < 1 \tag{3.29}$$

$$\min_{K_{12}\in S} \left\| \frac{-\dfrac{W_{perf}K_{12}G}{1+K_{11}G}}{\dfrac{W_{yc}K_{12}}{1+GK_{11}}} \right\|_{\infty} < 1 \tag{3.30}$$

S is the set of stable controllers. *Wperf* is a weighting function representing with a low-pass filter for the rejection of disturbance. *Wyc* is an another weighting function for reducing control effort. Further, *Wyc* ensures stability against the uncertainty in operating conditions.

3.4.3 Case study

The single-line diagram (SLD) of modified WF-integrated series-compensated 39-bus New England test system is shown in Figure 3.14. PMUs are available at every PCCs, i.e. bus-W_1, bus-W_2, and bus-W_3. A TCSC is connected at the middle of the transmission line 14-W_1. The lines 3-W_2 and 16-W_3 are series compensated

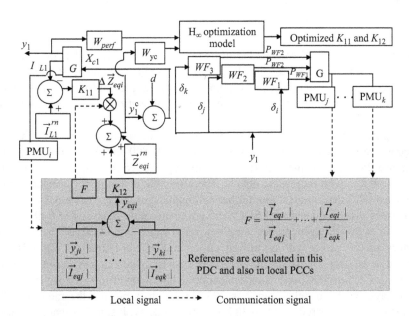

Figure 3.14   Robust SDC model for SSR damping at WF-connected power systems

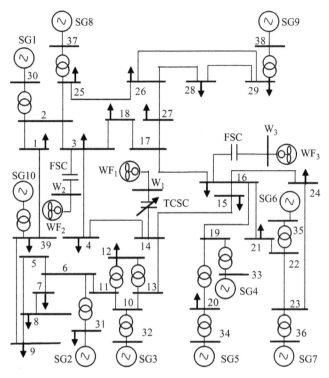

*Figure 3.15   Modified WF-integrated series-compensated 39-bus New England
test system*

with FSCs. Figure 3.15 shows SSO in line 14-W_1 for change in the compensation
level at the line from 25% to 45%. It is clear that the SDC (cumulative effect of K_{11}
and K_{12}) provides better damping as compared to local control (only K_{11}). Thereby,
synchrophasor technology is effective in SSO mitigation at the WF-connected
transmission networks.

   Figure 3.16 shows bode diagram of frequency responses of SDC with different
communication delays. Up to 0.1 Hz, even with 800 ms delay, the frequency
response does not change significantly. However, above this frequency, the magnitude reduces with increasing delays. Therefore, the damping performance of the
proposed SDC deteriorates with higher communication delay. Table 3.4 illustrates
gain cross over frequencies of SDC with communication delays. The 0 dB gain is
obtained at two different frequencies, $\omega gc1$ and $\omega gc2$. It is observed that the SDC
performance is stable within the frequency range of $\omega gc1$ to $\omega gc2$. $\omega gc1$ is considered to be very low frequency in power system dynamics and causes an insignificant effect on SDC with communication delays. However, $\omega gc2$ is sensitive to
the delays. With no delay, $\omega gc2$ is 293 Hz which is outside the range of SSO
frequencies. $\omega gc2$ is reduced to 6.38 Hz at 800 ms delay. It signifies that the margin
of stability of SDC degrades with increasing communication delays.

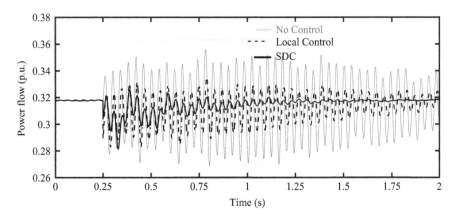

Figure 3.16   SSO in line 14-W₁ for change in compensation level at the line from 25% to 45%

Table 3.4   Gain cross over frequencies of SDC with communication delays

T_d (ms)	ω_{gc1}, ω_{gc1} (Hz)	τ (ms)	ω_{gc1}, ω_{gc1} (Hz)
0	0.016, 293	400	0.016, 12.8
100	0.016, 36.6	600	0.016, 8.65
200	0.016, 22.7	800	0.016, 6.4

3.5   SDC methodology to improve PFR in WF-integrated transmission systems

Structure of the SDC technique is classified as: (i) centralized control and (ii) distributed control. Figure 3.17 shows the structure of centralized SDC of grid-integrated WFs for improvement of PFR in transmission systems. A total n number of substations/PCCs, W1, W2,...,Wn, are identified for connecting WFs to transmission grid. Synchrophasor data at these substations are communicated at PDC for WAC action. In the centralized WAC approach, PDC uses system frequency, ($fW1$, $fW2$,..., fWn), calibrated with the PMUs at the WF-connected substations and the active power delivered by the WFs ($PW1$, $PW2$,..., PWn) to obtain power reference commands ($Pref1$, $Pref2$,..., $Prefn$). These reference commands are sent to the WAC for control action. The centralized WAC acknowledges data on estimated average rotor speed ($\omega1$, $\omega2$,..., ωn) and average wind speed ($vw1$, $vw2$,..., vwn) available at the WFs. Further, WAC communicates with PDC to receive the power references to generate WAC logics ($y_1^c, y_2^c,..., y_n^c$). The centralized controller realizes increased amount of data in proportion to the WF-connected high-voltage substations. Thereby, the complexity of the central WAC logic increases. Therefore, the computational time increases and added with communication delays between the centralized WAC and the

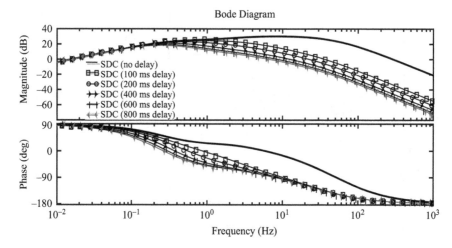

Figure 3.17   Bode diagram of frequency responses of SDC with different communication delays

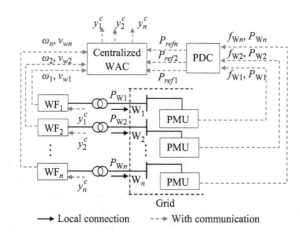

Figure 3.18   Structure of centralized SDC of grid-integrated WFs for improvement of PFR in transmission systems

PMUs at WF-connected high-voltage substations [45]. $y_1^c, y_2^c, \ldots, y_n^c$ are sent to the respective WFs (WF1, WF2 ,...,WFn) through different communication channels.

Figure 3.18 depicts the structure of distributed SDC of grid-integrated WFs for improvement of PFR in transmission systems. The active power references for the WFs is obtained in PDC in a similar way as discussed for centralized SDC. To obtain the distributed WAC logic, each local controller aggregates the estimated average rotor speed and average wind speed of their respective WF. Further, PDC communicates with the local controllers to send respective active power reference commands for each WF. Therefore, both the computational time and the complexity of distributed SDC logic are independent of the number of grid-connected WFs. Hence, distributed SDC techniques are simpler and more advantageous as compared to centralized SDC methods.

3.5.1 Synchronized P–f droop control of grid-integrated WFs using synchrophasor technology

The PDC receives WF active power injection to the transmission grid and system frequency, measured at the respective substations/PCCs with the PMUs. Further, PDC estimates available power reserve at each WF, which is used to obtain power share ratio (PSR) of the respective WF. The PSR information is sent to the local controller of the respective WF to compute synchronized P–f droop, as in (3.31). A detailed block diagram of such method is available in [46]. The synchronized P–f droop of *i*th WF is expressed as

$$R_i = \frac{R\Delta f_{Wi}}{a_i \sum_{i=1}^{n} \Delta f_{Wi}} \tag{3.31}$$

where Δf_{Wi} is the system frequency deviation observed at *i*th WF-connected high-voltage substation. R is the preset constant droop gain, and a_i is the PSR of *i*th WF. From the above expression, it is justified that the synchronized P–f droop gain is adaptive with grid frequency deviation at the PCCs.

3.5.2 Case study

The system for this case is shown in Figure 3.14. A 21 MW load is switched into bus-4 at 1 s. Figure 3.19 depicts system frequency oscillations at the WF-connected high-voltage substations for the event. With synchronized droop gain, the

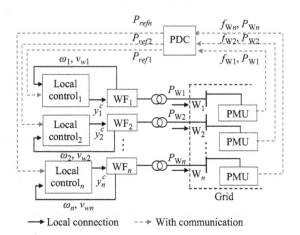

Figure 3.19   Structure of distributed SDC of grid-integrated WFs for improvement of PFR in transmission systems

oscillations settle faster than constant droop control. Figure 3.20 shows active power oscillations of the WFs for the load disturbance. It is observed that the oscillation damping is more with the synchronized droop gain as compared to the constant droop gain. About 5% data loss and random delays are introduced at the communication channel connected from PMU at bus-W1 to PDC. The delay distribution is 0.01–0.2 s with an average delay of 0.108 s and the standard deviation of the distribution is 0.055 s. Figure 3.21 depicts the effect of random communication delay distribution and data loss on system frequency at WF-connected high-voltage substations for load increase at bus-4. The results depict that the frequency is fluctuating at the PCCs which deteriorates PFR of the system. Further, the synchronized droop gain fails to damp out the frequency oscillations within 10 s, which

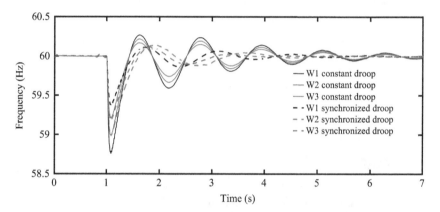

Figure 3.20   System frequency oscillations at the WF-connected high-voltage substations for the load disturbance

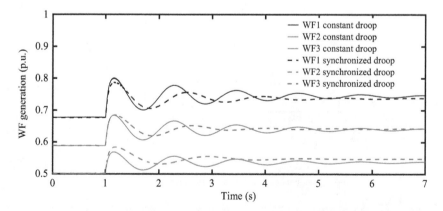

Figure 3.21   Active power oscillations of the WFs for the load disturbance

indicates that the SDC is vulnerable to communication delays and data drops. In such situation, WAPC technique can be used to compensate such issues.

3.6 Voltage oscillation damping in WF-integrated transmission network using synchrophasor technology

Synchrophasor technology has the potential to coordinate among WF-integrated high-voltage substations to obtain optimal deliverable reactive power injection from the WFs to the grid. The SDC method assists local droop control to make it variation of the droop gain in a synchronized way. Importantly, for this application, SDC uses synchrophasor data of PMUs connecting WF-connected substations only, to generate wide-area signals to PDC. Further, PDC communicates with the local controllers to execute synchronized Q–V droop gain at the WFs.

3.6.1 Synchronized Q–V droop control of grid-integrated WFs using synchrophasor technology

The SDC obtains optimal deviation in deliverable reactive power injection from jth WF as follows:

$$\Delta Q_j = \frac{\partial Q_j}{\partial V_j}\Delta V_j + \sum_{p=1}^{m}\frac{\partial Q_j}{\partial V_p}\Delta V_p + \frac{\partial Q_j}{\partial R_j}\Delta R_j + \Delta Q_j^{Loss}; p \neq j \tag{3.32}$$

$$\text{where,}\ \frac{\partial Q_j}{\partial V_j}\Delta V_j + \sum_{p=1}^{m}\frac{\partial Q_j}{\partial V_p}\Delta V_p = \Delta Q_j^{PDC} \tag{3.33}$$

m is the WF-connected high-voltage substations/PCCs employed in SDC. ΔV_p and ΔV_j are the voltage deviations at pth and jth PCC, respectively, w.r.t. nominal voltage. ΔR_j is the variation of synchronized droop gain R_j w.r.t. constant Q–V droop gain R.

Further, $Q_j^{Loss} = (I_j)^2 X_j \tag{3.34}$

Thereby, $\Delta Q_j^{Loss} = 2X_j I_j \Delta I_j + (I_j)^2 X_j \tag{3.35}$

I_j is the injected current from jth WF-connected substation to the transmission grid. ΔI_j is the deviation in the current injection due to ΔQ_j. X_j is the equivalent reactance of the low-/medium-voltage network in between jth WF and jth WF-connected substation. X_j can be estimated with ABCD parameter calculation for fixed topology in the off-line mode [47].

Substituting (3.33) and (3.35), in (3.32),

$$\Delta Q_j = \Delta Q_j^{PDC} + \frac{\partial Q_j}{\partial R_j}\Delta R_j + 2X_j I_j \Delta I_j + (I_j)^2 X_j \tag{3.36}$$

The PMUs at jth WF-integrated substation and other PCCs communicate with PDC to provide their respective voltage information and reactive power injection information. Thereafter, PDC forms

$$\begin{bmatrix} \Delta Q_1^{PDC} \\ \vdots \\ \Delta Q_m^{PDC} \end{bmatrix} = \begin{bmatrix} \dfrac{\partial Q_1}{\partial V_1} & \cdots & \dfrac{\partial Q_1}{\partial V_m} \\ \vdots & \ddots & \vdots \\ \dfrac{\partial Q_m}{\partial V_1} & \cdots & \dfrac{\partial Q_j}{\partial V_j} \end{bmatrix} \begin{bmatrix} \Delta V_1 \\ \vdots \\ \Delta V_m \end{bmatrix} \tag{3.37}$$

Further, PDC sends ΔQ_j^{PDC} to jth WF where $\frac{\partial Q_j}{\partial R_j}\Delta R_j$, $2X_jI_j\Delta I_j$, $(I_j)^2X_j$ are calculated with the local information to obtain ΔQ_j.

The synchronized Q–V droop at jth WF is obtained as

$$R_j = \frac{\Delta Q_j}{\Delta V_j} \tag{3.38}$$

3.6.2   Case study

SLD of 418-bus power network of South India is represented in Figure 3.22 [48]. Figure 3.22(a) shows the transmission system which constitutes of 78 substations of 400 kV or 220 kV level. Figure 3.22(b) depicts medium-/low-voltage network with 2.5 MW and 1.5 MW DFIG connections [49,50]. These DFIGs are represented in multiple groups, named as GA1, GA2, GA3, GB1, GB2, GC1, GC2, GC3, GC4, GD, GE, GF, GG, GH, and GI. Here each WF is

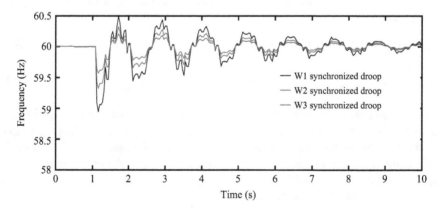

Figure 3.22   Effect of random communication delay distribution and data loss on system frequency at WF-connected high-voltage substations for load increase at bus-4

Table 3.5 Droop gain of DFIG groups

DFIG group	Average wind speed (m/s)	Variable droop gain (%)	Synchronized droop gain variation (% of variable droop)
GA1, GC1, GB2, GA3, GC4	12	9.37	−21.13 to 57.74
GB1, GA2, GC2, GC3	10	10.82	−31.70 to 36.60
GD	10	14.41	−51.35 to 2.71
GE	12	6.70	4.63–109.25
GF	12	17.65	−67.99 to −35.98
GG	14	4.90	15.31–13.06
GH	10	10.56	−16.48 to 67.05
GI	12	9.16	−3.71 to 92.58

represented with the DFIG groups which are integrated to a common 220 kV substation. In this way, the 418-bus system includes four WFs: W1 is connected to bus-71, W2 is connected to bus-36, W3 is connected to bus-51, and W4 is connected to bus-43.

Wind speed and droop gain at the DFIG groups are found from Table 3.5. The constant droop gain is made fixed at 10. The variable droop gain is calculated based on the availability of wind speed at the respective WFs [51]. However, for constant wind speed, the variable droop is fixed with time. On the other hand, synchronized droop gain varies with time even if the wind speed does not change. Thereby, the synchronized droop gain variation (in % of variable droop) for each group of DFIGs is provided in Table 3.5. A symmetrical fault is simulated at transmission line 39–33 at 500 ms, which causes tripling of the line at 600 ms. Figure 3.23 shows the voltages at the WF-connected high-voltage substations/PCCs for the disturbance. It is important to note that, even after the disturbance, voltages at the substations remain within the low-voltage ride through (LVRT) limit. Therefore, all the four WFs are connected to the grid during the oscillations. The oscillations damping with variable droop are more in comparison with constant droop. However, the variable droop is unable to perform better than SDC which damps the oscillations within 3 s, proving potential of the synchrophasor technology in this WAC application.

Further, in using SDC, impact delay distributions and 5% data loss on voltage disturbance at the WF-connected high-voltage substations are depicted in Figure 3.24. The delay distributions in communication channels among PMUs and PDC are provided in Table 3.6. Without WAPC, the SDC is unable to compensate such random delays and data loss. As a result, the voltages oscillate even after 5 s. The SDC with WAPC settle down the voltages within 3.5 s satisfactorily. This indicates the importance of compensating such delays and data loss in the PDC.

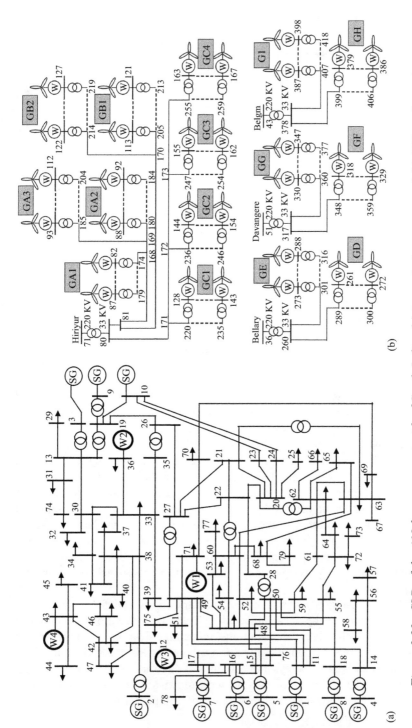

Figure 3.23 SLD of the 418-bus power network of South India: (a) transmission system and (b) WF configurations

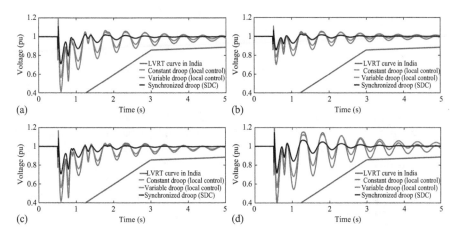

Figure 3.24   Voltage at (a) PCC-51, (b) PCC-43, (c) PCC-71, and (d) PCC-36, for line 39–33 tripping

Table 3.6   Delay distributions in communication channels among PMUs and PDC

	Maximum delay, minimum delay (ms)	Average delay (ms)	Standard deviation (ms)
PMU at W1-connected PCC/bus-71 to PDC	10, 100	72	24
PMU at W2-connected PCC/bus-36 to PDC	20, 200	168	65
PMU at W3-connected PCC/bus-51 to PDC	10, 160	124	43
PMU at W4-connected PCC/bus-43 to PDC	10, 60	40	18

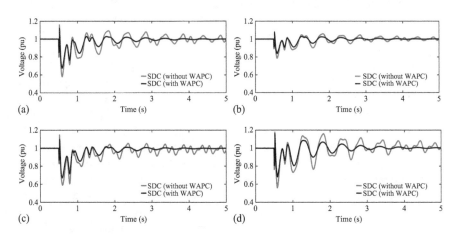

Figure 3.25   Impact of delay distributions (Table 3.6) and 5% data loss on voltage disturbances at (a) PCC-51, (b) PCC-43, (c) PCC-71, and (d) PCC-36, for line 39–33 tripping

3.7   Conclusions

From the above discussions, it is well realized that the synchrophasor technology has the potential in monitoring and controlling the power system even with variable and intermittent renewable resources. There are certain standards for accuracy level of PMUs that need to be followed up properly for effective WAC applications. The detailed frequency domain analysis of centralized WAC design is discussed to improve its robustness. Further, different droop-based applications of distributed control in WF-integrated power systems using synchrophasor data are presented. All these methods are proven their effectiveness as compared to local control techniques, in different dynamic situations of power transmission networks. However, such PMU-based control applications are vulnerable to significant communication delays and data drops, if not compensated. To overcome this issue, inclusion of the predictive controller in WAC framework is presented. It is noteworthy that the SDC techniques are formulated and incorporated with local controllers to improve the effectiveness of overall control performance. It is never recommended to use such WAC applications without support of existing local control approaches, due to the possibility of communication failure or insufficient bandwidth in the communication medium.

References

[1]   D. Dotta, A.S. Silva, and I.C. Decker, "Wide-area measurements based two-level control design considering signal transmission delay," *IEEE Trans. Power Syst.*, vol. 24, no. 1, pp. 208–216, 2009.

[2]   I. Kamwa and R. Grondin, "PMU configuration for system dynamic performance measurement in large multiarea power systems," *IEEE Trans. Power Syst.*, vol. 17, no. 2, pp. 385–394, 2002.

[3]   J. MacDowell, S. Dutta, M. Richwine, S. Achilles, and N. Miller, "Serving the future: advanced wind generation technology supports ancillary services," *IEEE Power Energy Mag.*, vol. 13, no. 6, pp. 22–30, 2015.

[4]   J. Tang, R. Achilles, B. Agrawal, *et al.*, "Reader's guide to subsynchronous resonance," *IEEE Trans. Power Syst.*, vol. 7, no. 1, pp. 150–157, 1992.

[5]   A.E. Leon and J.A. Solsona, "Sub-synchronous interaction damping control for DFIG wind turbines," *IEEE Trans. Power Syst.*, vol. 30, no. 1, pp. 419–428, 2015.

[6]   X. Zhang, X. Xie, H. Liu, and Y. Li, "Robust subsynchronous damping control to stabilise SSR in series-compensated wind power systems," *IET Gener. Transmiss. Distrib.*, vol. 13, no. 3, pp. 337–344, 2019.

[7]   H. Liu, X. Xie, X. Gao, H. Liu, and Y. Li, "Stability analysis of SSR in multiple wind farms connected to series-compensated systems using impedance network model," *IEEE Trans. Power Syst.*, vol. 33, no. 3, pp. 3118–3128, 2018.

[8]   X. Xie, W. Liu, H. Liu, Y. Du, and Y. Li, "A system-wide protection against unstable SSCI in series-compensated wind power systems," *IEEE Trans. Power Del.*, vol. 33, no. 6, pp. 3095–3104, 2018.

[9]   "Using synchrophasor data for oscillation detection," NASPI, Tech. Rep., October 2017. [Online]. Available: https://www.naspi.org/sites/default/files/reference_documents/crstt_oscillation_detection_20180129_final.pdf.

[10] S.O. Faried, I. Unal, D. Rai, and J. Mahseredjian, "Utilizing DFIG-based wind farms for damping subsynchronous resonance in nearby turbine-generators," *IEEE Trans. Power Syst.*, vol. 28, no. 1, pp. 452–459, 2013.

[11] B. Gao, R. Torquato, W. Xu, and W. Freitas, "Waveform-based method for fast and accurate identification of subsynchronous resonance events," *IEEE Trans. Power Syst.*, vol. 34, no. 5, pp. 3626–3636, 2019.

[12] L. Huang, H. Xin, Z. Li, P. Ju, H. Yuan, and G. Wang, "Identification of generalized short-circuit ratio for on-line stability monitoring of wind farms," *IEEE Trans. Power Syst.*, vol. 35, no. 4, pp. 3282–3285, 2020.

[13] E. Vittal, M.O. Malley, and A. Keane, "A steady-state voltage stability analysis of power systems with high penetrations of wind," *IEEE Trans. Power Syst.*, vol. 25, no. 1, pp. 433–442, 2010.

[14] R. Yousefian, R. Bhattarai, and S. Kamalasadan, "Transient stability enhancement of power grid with integrated wide area control of wind farms and synchronous generators," *IEEE Trans. Power Syst.*, vol. 32, no. 6, pp. 4818–4831, 2017.

[15] J. Liu and J. Cheng, "Online voltage security enhancement using voltage sensitivity-based coherent reactive power control in multi-area wind power generation systems," *IEEE Trans. Power Syst.*, vol. 36, no. 3, pp. 2729–2732, 2021.

[16] S. Wang, X. Meng, and T. Chen, "Wide-area control of power systems through delayed network communication," *IEEE Trans. Control Syst. Technol.*, vol. 20, no. 2, pp. 495–503, 2012.

[17] C. Huang, F. Li, T. Ding, Y. Jiang, J. Guo, and Y. Liu, "A bounded model of the communication delay for system integrity protection schemes," *IEEE Trans. Power Del.*, vol. 31, no. 4, pp. 1921–1933, 2016.

[18] *IEEE Standard for Synchrophasor Data Transfer for Power systems*, IEEE Std C37.118. 2-2011 (Rev. IEEE Std C37.118-2005), pp. 1–53, December 2011.

[19] J.W. Stahlhut, T.J. Browne, G.T. Heydt, and V. Vittal, "Latency viewed as a stochastic process and its impact on wide area power system control signal," *IEEE Trans. Power Syst.*, vol. 23, no. 1, pp. 84–91, 2008.

[20] F. Zhang, Y. Sun, L. Cheng, X. Li, J.H. Chow, and W. Zhao, "Measurement and modeling of delays in wide-area closed-loop control systems," *IEEE Trans. Power Syst.*, vol. 30, no. 1, pp. 2426–2433, 2015.

[21] W. Yao, L. Jiang, J. Wen, Q. Wu, and S. Cheng, "Wide-area damping controller for power system interarea oscillations: a networked predictive control approach," *IEEE Trans. Control Syst. Technol.*, vol. 23, no. 1, pp. 27–36, 2015.

[22] B. Naduvathuparambil, M. Valenti, and A. Feliachi, "Communication delays in wide area measurement systems," in *Proc. 34th Southeastern Symp. Syst. Theory*, 2002, pp. 118–122.

[23] Y. Zhao, G. Liu, and D. Rees, "Design of a packet-based control framework for networked control systems," *IEEE Trans. Control Syst. Technol.*, vol. 17, no. 4, pp. 859–865, 2009.

[24] B.P. Padhy, S.C. Srivastava, and N.K. Verma, "A wide-area damping controller considering network input and output delays and packet drop," *IEEE Trans. Power Syst.*, vol. 32, no. 1, pp. 166–176, 2017.

[25] J. Wu and T. Chen, "Design of networked control systems with packet dropouts," *IEEE Trans. Autom. Control*, vol. 52, no. 7, pp. 1314–1319, 2007.

[26] S. Skogestad and I. Postlethwaite, *Multivariable Feedback Control: Analysis and Design*, 2nd ed. New York, NY: Wiley, 2005.

[27] *Robust Control Toolbox Users Guide, MathWorks Inc.*, Natick, MA, 2001.

[28] W. Yao, L. Jiang, J. Wen, Q.H. Wu, and S. Cheng, "Wide-area damping controller of FACTS devices for inter-area oscillations considering communication time delays," *IEEE Trans. Power Syst.*, vol. 29, no. 1, pp. 318–329, 2014.

[29] P. Mahish, A.K. Pradhan, and A.K. Sinha, "Wide area predictive control of power system considering communication delay and data drops," *IEEE Trans. Ind. Informat.*, vol. 15, no. 6, pp. 3243–3253, 2019.

[30] W. Yao, L. Jiang, J. Wen, Q. Wu, and S. Cheng, "Wide-area damping controller for power system interarea oscillations: a networked predictive control approach," *IEEE Trans. Control Syst. Technol.*, vol. 23, no. 1, pp. 27–36, 2015.

[31] K. Narendra, D. Fedirchuk, R. Midence, *et al.*, "New microprocessor based relay to monitor and protect power systems against sub-harmonics," in *Proc. IEEE Elect. Power Energy Conf.*, 2011, pp. 438–443.

[32] L. Wang, X. Xie, Q. Jiang, H. Liu, Y. Li, and H. Liu, "Investigation of SSR in practical DFIG-based wind farms connected series compensated power system," *IEEE Trans. Power Syst.*, vol. 30, no. 5, pp. 2772–2779, 2015.

[33] IEEE Committee Report, "Reader's guide to subsynchronous resonance," *IEEE Trans. Power Syst.*, vol. 7, no. 1, pp. 150–157, 1992

[34] A.E. Leon and J.A. Solsona, "Sub-synchronous interaction damping control for DFIG wind turbines," *IEEE Trans. Power Syst.*, vol. 30, no. 1, pp. 419–428, 2015.

[35] G. Irwin, A. Jindal, and A. Isaacs, "Sub-synchronous control interactions between type 3 wind turbines and series compensated AC transmission systems," in *Proc. IEEE Power Energy Soc. General Meeting*, July 2011, pp. 1–6.

[36] H. Liu, X. Xie, Y. Li, H. Liu, and Y. Hu, "Mitigation of SSR by embedding subsynchronous notch filters into DFIG converter controllers," *IET Gener. Transmiss. Distrib.*, vol. 11, no. 11, pp. 2888–2896, 2017.

[37] J. Adams, C. Carter, and S.-H. Huang, "ERCOT experience with subsynchronous control interaction and proposed remediation," in *Proc. Transmiss. Distrib. Conf. Expo.*, 2012, pp. 1–5.

[38] X. Xie, X. Zhang, H. Liu, H. Liu, Y. Li, and C. Zhang, "Characteristic analysis of subsynchronous resonance in practical wind farms connected to

series-compensated transmissions," *IEEE Trans. Energy Convers.*, vol. 32, no. 3, pp. 1117–1126, 2017.

[39] X. Zhang, X. Xie, H. Liu, and Y. Li, "Robust subsynchronous damping control to stabilise SSR in series-compensated wind power systems," *IET Gener. Transmiss. Distrib.*, vol. 13, no. 3, pp. 337–344, 2019.

[40] H. Liu, X. Xie, X. Gao, H. Liu, and Y. Li, "Stability analysis of SSR in multiple wind farms connected to series-compensated systems using impedance network model," *IEEE Trans. Power Syst.*, vol. 33, no. 3, pp. 3118–3128, 2018.

[41] X. Xie, W. Liu, H. Liu, Y. Du, and Y. Li, "A system-wide protection against unstable SSCI in series-compensated wind power systems," *IEEE Trans. Power Del.*, vol. 33, no. 6, pp. 3095–3104, 2018.

[42] "Using Synchrophasor Data for Oscillation Detection," North American Synchro Phasor Initiative, October 2017. [Online]. Available: https://www.naspi.org/sites/default/files/reference_documents/crstt_oscillation_detection_20180129_final.pdf

[43] S.O. Faried, I. Unal, D. Rai, and J. Mahseredjian, "Utilizing DFIG-based wind farms for damping subsynchronous resonance in nearby turbine generators," *IEEE Trans. Power Syst.*, vol. 28, no. 1, pp. 452–459, 2013.

[44] P. Mahish, and A.K. Pradhan, "Mitigating subsynchronous resonance using synchrophasor data based control of wind farms," *IEEE Trans. Power Del.*, vol. 35, no. 1, pp. 364–376, 2020.

[45] Z. Wang and W. Wu, "Coordinated control method for DFIG-based wind farm to provide primary frequency regulation service," *IEEE Trans. Power. Syst.*, vol. 33, no. 3, pp. 2644–2659, 2018.

[46] P. Mahish and A.K. Pradhan, "Distributed synchronized control in grid integrated wind farms to improve primary frequency regulation," *IEEE Trans. Power Syst.*, vol. 35, no. 1, pp. 362–373, 2020.

[47] P. Mahish and S. Mishra, "Synchrophasor data based Q-V droop control of wind farm integrated power systems," *IEEE Trans. Power Syst.*, in early access.

[48] V.S.S. Kumar, "418 bus equivalent system of Indian southern grid," [Online]. Available: https://people.iith.ac.in/seshadri/Data/418-Bus-System-Data.pdf

[49] V.S.S. Kumar, "2.5 MW doubly fed induction generator," [Online]. Available: https://people.iith.ac.in/seshadri/Data/2p5-DFIG-Data.pdf

[50] V.S.S. Kumar, "1.5 MW doubly fed induction generator," [Online]. Available: https://people.iith.ac.in/seshadri/Data/1p5-DFIG-Data.pdf

[51] Y. Li, Z. Xu, Z. Zhang, and K. Meng, "Variable droop voltage control for wind farm," *IEEE Trans. Sustain. Energy*, vol. 9, no. 1, pp. 491–493, 2018.

Chapter 4

Power oscillation damping control for inter-area mode in reduced order model

Lalit Kumar[1], Nand Kishor[2] and Soumya R. Mohanty[3]

This chapter presents the study on design of local power system stabilizers (PSS) for the northern regional power grid (NRPG) system of India to control the oscillation due to least damped inter-area mode. First, eigenvalue analysis is conducted on linearized model of the NRPG system to obtain the properties of inter-area modes. The location of PSS in the network is suggested by applying residue technique. The power–voltage reference (*PVr*) characteristic of suggested generator is used to design the compensating block of local PSS. In order to make the PSS tuning fast after installation, the reduced order model (ROM) for the NRPG system is obtained. The same ROM has also been utilized in the initial design of local PSS by considering it in the tuning process of PSS-gain. The model order reduction (MOR) is performed with a target on preserving the important properties of the desired mode in ROM as good as possible. Validation of mode preservation in ROM is done by matching four information associated with the desired mode i.e. mode characteristics, modal observability, model controllability, and step response. Two local PSSs are designed for G44 and G15 of the NRPG system and their performance is investigated.

4.1 Introduction

For power system stability, practical rule is imposed for controller design which is to provide at least 5% damping to inter-area mode [1,2]. Inter-area oscillation (or mode) is one of the types in the category of electromechanical oscillations (or rotor modes) [3]. The concept behind the control of inter-area oscillation is to provide the additional damping torque via automatic voltage regulator of the generator [4]. The phase compensation based on classical control theory for such power system stabilizer (PSS) [5] has been effectively and widely adopted to damp out inter-area oscillation [6,7]. The authors in [8] studied and compared the methods for

[1]PEC Chandigarh, India
[2]Østfold University College, Norway
[3]Department of Electrical Engineering, IIT BHU, India

compensation. Two conventional techniques for designing the compensator part of PSS includes (i) generator excitation-power response (GEP) and (ii) power-voltage reference (PVr) characteristics [4,7]. Both techniques are widely adopted and compared for PSS design [6,7,9]. The transfer function (TF), $GEP(s)$ of the *i*th generator is the ratio of terminal voltage to the reference voltage [4] i.e. $GEP(s) = \frac{\triangle V_t(s)}{\triangle V_r(s)}$. And the TF, $PVr(s)$ (or PVr characteristic) of the *i*th generator is the ratio of "torque of electromagnetic origin" to the reference voltage of AVR [4]. The author in [10] modified the GEP TF approach and has shown that phase and magnitude of modified GEP TF is in close approximation with that of PVr TF. It is mentioned in [4] that TF, $PVr(s)$ is almost invariant for a wide range of operating conditions, and thus PSS with fixed parameters tend to be robust.

Though the subject of wide-area PSS has attracted many researchers in the past decade, but there are some related issues, the foremost is time-delay [11,12]. In [12], the authors utilized a recently reported benchmark "JAYA algorithm" [13] to tune the wide-area PSS. A time-delay compensator is designed for variable communication time-delays. On the other hand, if the delays in remote signals are not appropriately compensated, the performance of PSS may get deteriorated [12]. Further, the authors insist on the need for cyber-attack resilient system for remote data transfer in the future scope. Being more sensitive to changing operating conditions, wide-area PSS requires more frequent tuning than local PSS. Some other issues associated with wide-area PSS include (i) cost, (ii) less reliability, and (iii) require close monitoring system. Due to these reasons, many authors [10,14] preferred the use of local PSS over wide-area PSS. Sometimes, the damping target equivalent to wide-area PSS can be achieved by placing two local PSS at the far end of the interconnected tie-line.

The authors in [7] observed that the left-shift in low-frequency modes using power-PSS are greater than speed PSS. Residue approach [6] is a conventional approach in the PSS design. In [15], improved residue matrix is applied for optimal placement. The study [14] has provided full-ranked optimal conditions for optimal placement of PSS.

To deal with computational burden in PSS design for large-scale power systems (LSPSs), researchers in their methodology often adopt two alternatives i.e. model order reduction (MOR) and intelligent optimization techniques for tuning [12,15,16]. Sometimes, both of these alternatives are utilized for LSPSs [14,17]. Some authors avoid the computational burden with study limited to low order Heffron Phillip's model [18] of single machine infinite bus (SMIB) system for PSS design. In [19], the PSS is tuned by solving a delicately formulated optimization problem with the help of proposed differential evolution algorithm.

Many forms of PSSs have been proposed and compared over the years having different structures and design methodology [15,20,21]. The authors in [21] studied the four types of PSS structures: conventional PSS, single neuron-based PSS (SNPSS), adaptive PSS (APSS), and multi-band PSS. Similarly, the authors in [17] have compared the proposed PSS with multi-band PSS4B, conventional $\triangle \omega$ PSS, conventional $\triangle P_a$ PSS, and particle swarm optimization-based PSS. In [10], a PSS design method is proposed by modifying the linearized procedure wherein the external reactance evolving from step-up transformer is considered in the linearized dynamics. In [14], the authors proposed integrated architecture for optimal placement and PSS tuning. In [22], the PSS is designed for 54th unit of northern regional power grid (NRPG) system by incorporating the model order reduction. The electromechanical mode damping is determined to be 0.006323 after tuning of PSS in original and reduced order model. However, the mode of oscillation corresponding to said damping is not referred in their study. Also, no information can be found on damping before tuning of PSS in the system.

This chapter provides the study on the design of power-PSSs to control the least damped inter-area mode in the NRPG system. The residue approach is utilized to identify the suitable generator location for placement of local power-PSS. For designing compensating block of power-PSS for the located generator, *PVr* characteristic of the generator is determined. The reduced order model (ROM) is obtained with a target on preserving the least damped inter-area mode in ROM. The performance of the designed PSSs is investigated by the case study which covers the practicality of the wide-area disturbances.

The remainder of the chapter is organized as follows. Section 4.2 presents the eigenvalue analysis of the NRPG system. Section 4.3 reviews the residue approach, the basic of conventional PSS and the compensation method associated with PSS design. Section 4.4 presents the model order reduction of the NRPG system. PSS design for the NRPG system is discussed in Section 4.5. Section 4.6 investigates the performance of PSS44 and PSS15 under the case study. At last, the conclusions are drawn in Section 4.7.

4.2 Small signal stability: eigenvalue analysis of NRPG system

The NRPG system of India is the largest among all five regional power grids in India, which covers around 30% geographical area and 28% population and contributes to nine states of India. The modeling data of NRPG [22] system consists 60 machines, 246 buses, 376 lines, 44 exciters, and 30 static VAR compensators (SVCs). No PSS and thyristor controlled series capacitor (TCSC) are installed on any of the generator. Among 246 busses, 46 buses are the high voltage buses of 400 kV. Since, only 44 generators are considered to have exciters therefore, so these have the possibility of local PSS placement. Power system toolbox (PST) [23] is used to obtain the linearized state space model of the NRPG system. The full linearized system obtained is of order 482. The *i*th complex eigenvalue (mode) in the linearized system can be represented as

$$\lambda_i = \sigma_i \pm j\omega_{d_i} \tag{4.1}$$

where $\sigma_i = -\zeta_i \omega_{ni}$ and $\omega_{d_i} = \omega_{ni}\sqrt{1-\zeta_i^2}$ and damping ratio, ζ and frequency, f_n can be obtained as follows:

$$\zeta_i = -\frac{\sigma_i}{\sqrt{\sigma_i^2 + \omega_{d_i}^2}} \quad \text{and} \quad f_{n_i} = \frac{\omega_{n_i}}{2\pi} \tag{4.2}$$

Table 4.1   Characteristics of inter-area modes in NRPG system

System mode no.	Inter-area mode no.	Damping ratio (ζ)	Natural frequency in Hz (f_n)
132	I	0.0327	0.768
125	II	0.0894	0.730

Two inter-area modes are found in the system by eigenvalue analysis, which are characterized in Table 4.1. As can be seen in Table 4.1, that inter-area mode I has damping lower than 5%, therefore, the control design will be focused on said mode, which will be referred as a desired mode in the rest of the chapter.

4.3 Review on residue approach, conventional PSS and compensation associated with PSS design

4.3.1 Residue approach for location of PSS

As the PSS is output feedback controller, so its location should also be based on measurable output signal for feedback. Therefore, the input/output-oriented approach called "residue approach" for the location PSS is introduced.

For the general linearized state space model of the power system having p-outputs q-inputs of order, n, the TF, $G(s)$ can be written as [6,24]:

$$G(s) = C(sI - A)^{-1}B = \sum_{i=1}^{i=n} (Cu_i) \bullet \frac{(v_i'B)}{s-\lambda_i} = \sum_{i=1}^{i=n} \frac{R_i}{s-\lambda_i} \tag{4.3}$$

where Cu_i is known as the modal/output observability and $v_i'B$ is known as the modal/input controllability of ith mode and being the product of modal observability and modal controllability, R_i is known as the modal residue of ith mode. It may be seen in (4.3) that the appearance of modes in $G(s)$ with poor damping i.e. $|real(\lambda_i)| \approx 0$ depends on the residue, R_i. And that is why the modal controllability and the modal observability of the system are to be determined for the control location of ith mode, λ_i. Unlike participation factor matrix [5], the residue matrix is not a square matrix and its dimension depends on the number of inputs, p and outputs, q i.e. $[p \times q]$. The maximum element in R_i i.e. R_{ikl}^{max} indicates that the kth output signal (controlling signal) should be feedback to lth input (control input) through the controller for maximum damping of ith mode [24]. For local controller, these indices, k and l, should belong to the same generator for which the element R_{ikl} (local residue) may not be the maximum value in R_i.

4.3.2 Conventional PSS: basics

PSS, being the output feedback control, requires a feedback loop from output of the system to input of the system through a PSS TF block. The TF of the ith PSS can be written as [5]

$$PSSi(s) = K_i \left(\frac{sT_{wi}}{1 + sT_{wi}} \right) \left(\frac{1 + sT_{1i}}{1 + sT_{2i}} \right) \left(\frac{1 + sT_{3i}}{1 + sT_{4i}} \right) = K_i \bullet G_{wi}(s) \bullet G_{ci}(s) \quad (4.4)$$

where K_i is the PSS-gain, $G_{wi}(s)$ is the wash out filter to attenuate the frequencies lower than the inter-area modes. The cut-off frequency in wash out filter is generally chosen as 0.1 Hz [6] which is also followed in this chapter. $G_{ci}(s)$ is the phase compensating block which will be discussed later in detail.

The output of the PSS is a voltage signal (control input) which is fed to the exciter's reference voltage of generator. PSS takes the speed state as an input and thus is called as speed PSS. Speed state can also be obtained by generator's electric power, P_e and its inertia, H as: $\Delta\omega = \frac{-\Delta P_e(s)}{2Hs}$ and thus the PSS can also take ΔP_e as an input by adding a simple block, $\frac{-1}{2Hs}$ to speed PSS. Such PSSs are called power-PSSs.

More details on conventional PSS and its modeling can be followed from Refs. [5,6]. In this chapter, the design of power-PSS is preferred against speed PSS, being referred as PSS in the rest of the chapter. Two important steps in PSS design include: (i) design of the compensation block, $G_c(s)$ and (ii) tuning of PSS-gain, K.

4.3.3   Compensation using PVr characteristics

In this study, *PVr* characteristic approach is preferred for design of compensating block in PSS. As discussed in Section 4.1, the *PVr* characteristic of *i*th generator is the ratio of "torque of electromagnetic origin" to the reference voltage of AVR. The torque of electromagnetic origin is a virtual parameter, which is not reflected in any system variable directly. It is computed indirectly as the electrical power output of the generator after disabling the shaft dynamics (or rotor modes) [4]. In small signal stability analysis, the per-unit perturbations in electrical power and electrical torque are identical. According to the definition, the TF, *PVr*(s) can be written as:

$$PVr(s) = \frac{\Delta P_e(s)}{\Delta V_r(s)} \Bigg|_{shaft \ dynamics \ disabled} \quad (4.5)$$

where $\triangle P_e$ and $\triangle V_r$ are the perturbation in electrical torque and reference voltage, respectively. Next paragraph discusses the role of *PVr*(s) in the design of PSS's compensator.

As mentioned in Section 4.1, the objective of PSS design is to add the torque proportional to machine speed (damping torque) on the generator shaft over the frequency range of the rotor modes [4] i.e. [0.01 2] Hz. This objective can be represented mathematically by introducing a proportionality constant called "damping torque coefficient", $D(s)$ given by:

$$D(s) = \frac{\Delta P_e(s)}{\Delta\omega(s)} \Bigg|_{[0.1 \ 2] \ Hz} = \frac{\Delta P_e(s)}{\Delta V_r(s)} \bullet \frac{\Delta V_r(s)}{\Delta\omega(s)} \Bigg|_{[0.1 \ 2] \ Hz} \quad (4.6)$$

where $\frac{\Delta P_e(s)}{\Delta V_r(s)}$ is the TF between the electrical torque (power) and the AVR reference and $\frac{\Delta V_r(s)}{\Delta\omega(s)}$ is TF between the AVR reference and the machine speed. Referring (4.6), for pure damping, $D(s)$ should be a real number (p.u. on machine base). To achieve this task, a controller with TF, $\frac{\Delta V_r(s)}{\Delta \omega(s)}$ can be designed which could compensate the phase of generator TF, $\frac{\Delta P_e(s)}{\Delta V_r(s)}$ in the range of rotor modes. Referring (4.6), in order to achieve the controller's TF, $\frac{\Delta V_r(s)}{\Delta \omega(s)}$ free from the rotor modes, the generator TF, $\frac{\Delta P_e(s)}{\Delta V_r(s)}$ must be made free from the same. Thus, generator TF, $\frac{\Delta P_e(s)}{\Delta V_r(s)}$ is computed after disabling the shaft dynamics (rotor modes). The shaft dynamics of all the machines can be disabled in linearized state space model by simply deleting the rows and/or columns corresponding to "rotor angle" and "speed" states from ABCD matrices [4,7]. The resulting TF is called *PVr* characteristics as given in (4.5). The form of controller with TF as $\frac{\Delta V_r(s)}{\Delta \omega(s)}$ is generally known as PSS. Thus, (4.6) after disabling shaft dynamics can be rewritten as

$$D(s) = PVr(s) \bullet PSS(s)|_{[0.1\ 2]\ \text{Hz}} \tag{4.7}$$

where $PSS(s)$ is the TF of the PSS for the same generator as the *PVr* characteristics was for. Referring (4.7), in order to bring $D(s)$ a real number, compensating block, $G_{ci}(s)$ of PSS should satisfy (4.8) given below [4,7]:

$$phase\left\{ G_{ci}^{\uparrow\downarrow}(s)PVr(s) \right\} \bigg|_{[0.1\ 2]\text{Hz}} \approx \pm 2n\pi$$

$$\text{or} \quad phase\left\{ \frac{\left(1 + T_1^{\uparrow\downarrow}s\right)\left(1 + T_3^{\uparrow\downarrow}s\right)}{\left(1 + T_2^{\uparrow\downarrow}s\right)\left(1 + T_4^{\uparrow\downarrow}s\right)} PVr(s) \right\} \Bigg|_{[0.1\ 2]\text{Hz}} \approx \pm 2n\pi \quad n = 0, 1, 2 \ldots$$

$$\tag{4.8}$$

where the superscript "↑↓" on $G_{ci}(s)$ and $T_j, j = 1, 2, 3, 4$ signifies the tuning. The parameters, T_js of compensator needs to be tuned on the phase plot of loop TF i.e. $G_{ci}(s) \bullet PVr_i(s)$ to satisfy (4.8) approximately.

The order of $PVr(s)$ is always significantly lower than original system order due to disabling of the shaft dynamics. But, for LSPS, the order of $PVr(s)$ may still be high enough which causes the initial tuning of $G_{ci}(s)$ (new PSS) to be challenging and time consuming. Even the optimization techniques may take minutes to hours to give optimal solution for $G_{ci}(s)$. It is known that any optimization algorithm converges fast only if the starting solution (initial assumption) is close to the final solution. The initial designing of any new PSS has to go through the time-consuming process for LSPS. As discussed in Section 4.1, once the new PSS is designed, the role of optimization technique and the model order reduction come into play for fast tuning of the PSS.

4.4   Model order reduction of NRPG system

It is well known that performance of MOR methods varies with system complexities. In such situation, the system needs to be tested with different MOR methods, while retaining the desired mode properties. The objective is to obtain the lowest possible reduced order model (ROM) from the full-order NRPG system such that the desired mode gets preserved. In this process, following MOR techniques are tested:

(i)   *Method-I:* Singular value decomposition (HSVD)-based state elimination with restricting the Gramian to frequency interval, [0.012] Hz
(ii) *Method-II:* Truncation
(iii) *Method-III:* Mode separation with cut off frequency, 2 Hz
(iv) *Method-IV:* Mode separation with cut off frequency, 1 Hz
(v)   *Method-V:* State elimination

The related MATLAB® function used in MOR and the corresponding lowest satisfactory order are listed in Table 4.2. The properties of the desired mode obtained in ROMs are given in Table 4.3, which can be compared with those of original system (see Table 4.1).

It can be seen in Table 4.3 that although 40th-order ROM using "truncation" method is capable to preserve the oscillation frequency of desired mode but, alters the damping of the mode. On the other hand, remaining MOR methods though preserve the desired mode but are of comparatively larger order size. The active power response of G39 (TANDA2), being the highest observable generator in the desired mode, is shown in Figure 4.1 for the original system and the ROMs. Based on Table 4.3 and well-matched response in Figure 4.1, it can be said that the first method (see Table 4.2) provides the satisfactory lowest order ROM. In the remaining discussion, ROM will be referred to the 60th-order model. Further, more validation on the preservation of mode properties is analyzed. The modal controllability of desired mode is shown in Figure 4.2, for both, full-order system and

Table 4.2   Tested MOR techniques

S. no.	MOR technique	Related MATLAB function	Satisfactory order
1.	Method-I	Balred with 'HSVD' option	60
2.	Method-II	Balred with 'Truncate' option	40
3.	Method-III	modsep	435
4.	Method-IV	modsep	144
5.	Method-V	Balred with 'StateElimMethod' option	135

Table 4.3   Properties of the desired mode in ROMs

MOR method no. in Table 4.2	Damping of preserved desired mode	Frequency of preserved desired mode
I	0.034	0.768
II	0.029	0.767
III	0.033	0.768
IV	0.033	0.768
V	0.041	0.768

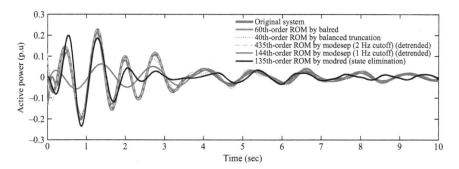

Figure 4.1   Response of G39 (TANDA2) subjected to 10% step input to exciter of G44

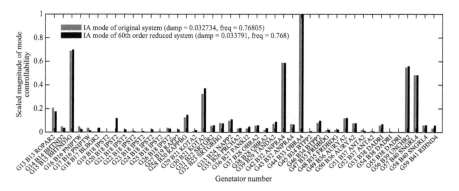

Figure 4.2   Bar chart for modal controllability of desired inter-area mode for full and reduced systems

ROM. It can be seen that modal controllability property is very well retained in ROM. Similarly, the modal observability of desired mode is shown in Figure 4.3 for both, full system and ROM. Only ten major observable generators are shown in the said figure. It is clear that modal observability property is also very well pre-served in ROM. Thus, these comparisons suggest that the obtained ROM has retained the desired mode properties and can be used for the design of PSS.

4.5   PSS design for NRPG system

The complete methodology for PSS design utilizing the MOR is illustrated in Figure 4.4. The residue approach discussed in Section 4.3.1 is applied on the NRPG system and the residue matrix, R_{132} of dimension (60×44) is obtained for the desired mode. Since only 44 generators have exciters therefore, only 44 local residues can be found in R_{132}. Among these 44 local residues, the two highest local residues are found to be associated with G44 (OBRA4) and G15 (BHTND3G),

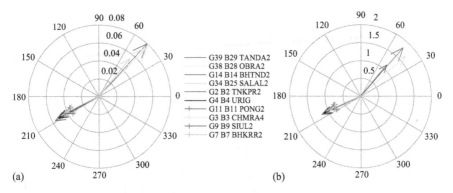

Figure 4.3 Compass plot for modal observability of desired inter-area mode for (a) full system and (b) ROM

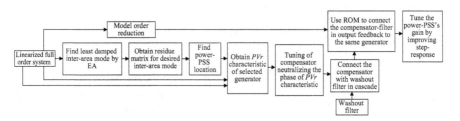

Figure 4.4 Complete methodology of local PSS design incorporating MOR

which suggests the best location for PSSs installation. Though the ROM of 60th order is a well approximate for specific task, but it cannot be utilized to tune the compensator block of PSS based on the *PVr* characteristic.

The reason lies in the fact that the shaft dynamics cannot be disabled in ROM because of loss of internal states in ROM. In the given system with 60 generators, shaft dynamic states, i.e. rotor angle states, $\Delta\delta$ and speed states, $\Delta\omega$ account to 120 in numbers. However, the obtained ROM itself is the order of 60 only. So, 120 shaft dynamic states cannot be disabled in 60th-order ROM. In addition, ROM does not have any information about 60 states in it. Only the input/output information is known in the ROM. But, the ROM can be utilized to tune the PSS-gain and thus MOR allows the fast and precise tuning for optimal gain. G44, being associated with top highest residue, will be preferred for PSS design. The *PVr* characteristic of G44 is obtained in the form of TF, $PVr_{44}(s)$. Two poles and two zeros of $G_{c44}(s)$ are tuned in "control system designer" of MATLAB [25], to satisfy (4.8). The option of optimization-based tuning method [26] is chosen with the initial guess of 0.1 for all T_js of the compensator. The compensation TF, $G_{c44}(s)$

obtained as such is

$$G_{c44}(s) = \frac{(1 + 0.224s)}{(1 + 8.019s)} \frac{(1 + 18.49s)}{(1 + 0s)}$$ (4.9)

When the power-PSS TF is written in the form of time constant, the gain of conventional power-PSS can be described as $K_{p-PSS} = (-1/2H \times K \times T_w)$. This gain is also tuned following the same steps and is obtained as -9.5889. As a common practice, one-third of K^c_{p-PSS} is selected as power-PSS. Now, the complete TF of local power-PSS for G44 is obtained as

$$PSS44(s) = \frac{-3.1963(1 + 0.224s)(1 + 18.49s)}{(1 + 10s)(1 + 8.019s)(1 + 1.66 \times 10^{-06}s)}$$ (4.10)

The next highest local residue is associated with G15 (BHTND3G) for the desired mode. Following the same above PSS design steps, PSS15 is designed for the same desired mode, which is obtained as

$$PSS15(s) = \frac{-6(1 + 0.1275s)(1 + 1.763s)}{(1 + 10s)(1 + 0.766s)(1 + 1.62 \times 10^{-06}s)}$$ (4.11)

To mention here, authors [14] also designed the PSS for the NRPG system, placed on G54 (DADRI). But, no information is given about inter-area modes in the system, and for which mode the controller has been designed. Further, in their study, 15th-order ROM has not been validated for the preservation of desired mode. On the other hand, the study presented here discussed it before discussing the local-PSS design. In the next section, the performance of $PSS44(s)$ and $PSS15(s)$ is investigated.

4.6   Case study for investigating PSS44 and PSS15 performance

As mentioned in Section 4.1, the desired damping equivalent to wide-area PSS can be achieved by placing two local PSS at far end of interconnected tie line. Here, the case study will be focused on heavily loaded tie lines. The three heavily loaded 400 kV lines in the NRPG system are given in Table 4.4, out of which, two are comparably longer than the third (line 376). Two distant groups of generators are located at a distance of 1,010 km as indicated on the geographical map [27] shown in Figure 4.5. Each group is referred by the name of generator having maximum local residue, i.e. OBRA4 group and BHTND3G group. The report [28] also discussed about the severity of these three lines mentioning that the short circuit levels at DADRI, BAWANA, and MANDLA are very high, and thus seek due attention for the investigation on the performance of designed PSSs. The step disturbance (10%) is fed to the exciter of G15 at 1 s and the deviation in active power flow (p. u.) on these lines are shown in Figure 4.6. The said figure illustrates the active power flow; without PSS, with PSS on G44 (PSS44) only, and with PSSs on both generators G44 & G15 (PSS44 and PSS15). It is indicated that response without

Table 4.4 Three heavily loaded lines in descending order of power flow

Line no.	From bus	To bus	Power flow in MW
333	235 DADRI4	240 MANDLA	1,545.9
376	235 DADRI4	38 DADRI	−1,200
187	154 BAWANA	240 MANDLA	−1,104.21

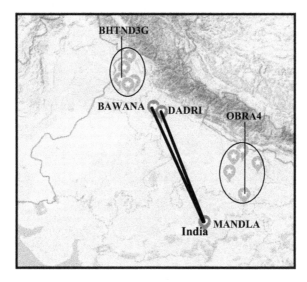

Figure 4.5   Geographical location of the studied group of generators and the tie lines [27]

PSS undergoes undue oscillation, while using PSS on G44 only, the oscillation magnitude is comparatively reduced. On the other hand, installing PSSs on both generators, G44 and G15, leads to further reduction in the magnitude of oscillation with a quick damping. Further to mention, a significant damping in oscillatory response of power flow can be observed on all the three heavily loaded tie lines.

The damping of 0.768 Hz mode is again analyzed by conducting eigenvalue analysis on the NRPG system under three cases of PSSs installation i.e. without PSS, with PSS on G44, and lastly PSSs on G44 and G15. And, the result for damping is shown in Figure 4.7 where it can be seen that, with PSS on G44 (PSS44), the damping of desired mode (0.768 Hz) has improved i.e. "more than 5%." The damping has further improved using PSS on both generators, i.e. PSS44 and PSS15. It should be noticed that, in addition to improvement in damping of the desired mode, the nearby inter-area mode (mode II in Table 4.1) also undergoes damping improvement.

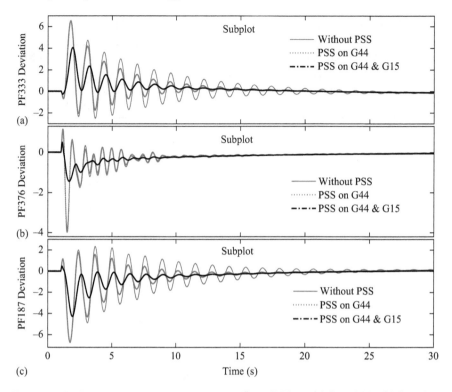

Figure 4.6   Deviation in p.u. active power flow (PF) in (a) line 333, (b) line 376, and (c) line 187

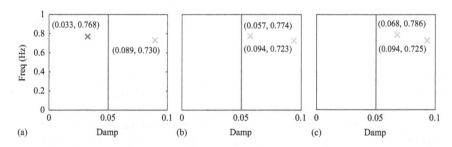

Figure 4.7   Inter-area mode properties (a) without PSS, (b) PSS on G44 (OBRA4), and (c) PSS on G44 and G15 (BHTND3G)

4.7 Conclusions

This chapter presented the study on PSSs design for least damped inter-area mode in the NRPG system. The eigenvalue analysis of the NRPG system gives 0.768 Hz mode as a least-damped inter-area mode. The optimal locations of local power-

PSSs are suggested by residue approach. The *PVr* characteristics of the located generators are used to design the compensating blocks of local power-PSSs. For fast PSS tuning, the MOR is performed targeting the desired mode to be preserved in ROM. The lowest satisfactory order of ROM is found to be as 60. Two PSSs are designed for G44 and G15. The performance of the designed PSSs is investigated by a case study. The results show that the designed PSSs perform satisfactorily to control the least-damped inter-area mode.

References

[1] S. Mendoza-Armenta and I. Dobson, "Applying a formula for generator redispatch to damp interarea oscillations using synchrophasors," *IEEE Transactions on Power Systems*, vol. 31, no. 4, pp. 3119–3128, 2016.

[2] C. Y. Chung, L. Wang, F. Howell, and P. Kundur, "Generation rescheduling methods to improve power transfer capability constrained by small-signal stability," *IEEE Transactions on Power Systems*, vol. 19, no. 1, pp. 524–530, 2004.

[3] L. Kumar and N. Kishor, "Local PSS design for inter-area oscillation in northern regional power grid of India," in *2017 IEEE PES Asia-Pacific Power and Energy Engineering Conference (APPEEC)*, Bangalore, India, 8–10 November 2017, pp. 1–6.

[4] M. Gibbard and D. Vowles, "Simplified 14-Generator Model of the South East Australian Power System," University of Adelaide, South Australia, 2014.

[5] P. Kundur, *Power System Stability and Control*. New York, NY: McGraw-Hill, Inc., 1994.

[6] B. C. Pal, *Robust Control in Power Systems*. New York, NY: Springer, 2005.

[7] M. J. Gibbard, P. Pourbeik, and D. J. Vowles, *Small-Signal Stability, Control and Dynamic Performance of Power Systems*. Adelaide: University of Adelaide Press, 2015.

[8] M. J. Gibbard and D. J. Vowles, "Reconciliation of methods of compensation for PSSs in multimachine systems," *IEEE Transactions on Power Systems*, vol. 19, no. 1, pp. 463–472, 2004.

[9] L. Kumar, N. Kishor, and Shweta, "Frequency monitoring of forced oscillation in PMU's data from NASPI," in *Proceedings of the 18th Mediterranean Electrotechnical Conference, Limassol, Cyprus, 18–20 April*, 2016, pp. 18–20.

[10] A. Kumar, "Power system stabilizers design for multimachine power systems using local measurements," *IEEE Transactions on Power Systems*, vol. 31, no. 3, pp. 2163–2171, 2016.

[11] S. Ghosh, K. A. Folly, and A. Patel, "Synchronized versus non-synchronized feedback for speed-based wide-area PSS: effect of time-delay," *IEEE Transactions on Smart Grid*, vol. 9, no. 5, pp. 3976–3985, 2018.

[12] T. Prakash, V. P. Singh, and S. R. Mohanty, "A synchrophasor measurement based wide-area power system stabilizer design for inter-area oscillation

damping considering variable time-delays," *International Journal of Electrical Power & Energy Systems*, vol. 105, pp. 131–141, 2019.

[13]   R. Venkata Rao, "Jaya: a simple and new optimization algorithm for solving constrained and unconstrained optimization problems," *International Journal of Industrial Engineering Computations*, vol. 7, pp. 19–34, 2016.

[14]   N. Kulkarni, S. Kamalasadan, and S. Ghosh, "An integrated method for optimal placement and tuning of a power system stabilizer based on full controllability index and generator participation," *IEEE Transactions on Industry Applications*, vol. 51, no. 5, pp. 4201–4211, 2015.

[15]   B. Ehsan-Maleki, P. Naderi, and H. Beiranvand, "A novel 2-stage WAPSS design method to improve inter-area mode damping in power systems," *International Transactions on Electrical Energy Systems*, vol. 28, no. 3, p. e2503, 2018.

[16]   M. A. Masrob and M. A. Rahman, "Design of a simplified fuzzy logic power system stabilizer for dynamic reduction of a power system model," in *IET International Conference on Resilience of Transmission and Distribution Networks (RTDN 2017)*, 2017, p. 5 (6.)-5 (6.).

[17]   A. Yaghooti, M. Oloomi Buygi, and M. H. Modir Shanechi, "Designing coordinated power system stabilizers: a reference model based controller design," *IEEE Transactions on Power Systems*, vol. 31, no. 4, pp. 2914–2924, 2016.

[18]   W. G. Heffron and R. A. Phillips, "Effect of a modern amplidyne voltage regulator on underexcited operation of large turbine generators [includes discussion]," *Transactions of the American Institute of Electrical Engineers. Part III: Power Apparatus and Systems*, vol. 71, no. 3, pp. 692–697, 1952.

[19]   D. Ke and C. Y. Chung, "Design of probabilistically-robust wide-area power system stabilizers to suppress inter-area oscillations of wind integrated power systems," *IEEE Transactions on Power Systems*, vol. 31, no. 6, pp. 4297–4309, 2016.

[20]   M. Farahani, "A multi-objective power system stabilizer," *IEEE Transactions on Power Systems*, vol. 28, no. 3, pp. 2700–2707, 2013.

[21]   P. He, F. Wen, G. Ledwich, Y. Xue, and K. Wang, "Effects of various power system stabilizers on improving power system dynamic performance," *International Journal of Electrical Power & Energy Systems*, vol. 46, pp. 175–183, 2013.

[22]   "Northern Regional Power Grid (NRPG) Data," *IIT Kanpur, Department of Electrical Engineering*. [Online]. Available: https://www.iitk.ac.in/eeold/facil-ities/Research_labs/Power_System/NRPG-DATA.pdf. Accessed: 21 September 2015.

[23]   "Power System Toolbox Webpage." [Online]. Available: http://www.eps.ee.kth.se/personal/vanfretti/pst/Power_System_Toolbox_Webpage/PST.html. Accessed: 14 September 2015.

[24]   Y. Yuan, Y. Sun, and L. Cheng, "Determination of wide-area PSS locations and feedback signals using improved residue matrices," in *APCCAS 2008 – 2008*

IEEE Asia Pacific Conference on Circuits and Systems, Macao, China, 30 November–30 December, 2008, pp. 762–765.

[25] "Getting Started with the Control System Designer – MATLAB & Simulink Example – MathWorks India." [Online]. Available: https://in.mathworks.com/help/control/examples/getting-started-with-the-control-system-designer.html. Accessed: 10 February 2017.

[26] "Optimize LTI System to Meet Frequency-Domain Requirements – MATLAB & Simulink – MathWorks India." [Online]. Available: https://in.mathworks.com/help/sldo/ug/optimize-lti-system-to-meet-frequency-domain-requirements.html#brzturj-5. Accessed: 14 December 2018.

[27] "Google Maps." [Online]. Available: www.google.co.in/maps/. Accessed: 08 December 2018.

[28] "Agenda note for 32nd Meeting of the Standing Committee on Power System Planning in Northern Region." [Online]. Available: https://docs.google.com/viewer?url=http%3A%2F%2Fwww.cea.nic.in%2Freports%2Fcommittee%2Fscm%2Fnr%2Fagenda_note%2F32nd.pdf. Accessed: 12 December 2018.

Chapter 5

Real-time congestion management in transmission networks

*Faheem Ul Haq[1], Kotakonda Chakravarthi[1]
and Pratyasa Bhui[1]*

The transmission network serves as the backbone of any power system. The possibility of power transportation from generation to consumption is done with the help of the transmission and distribution networks. Though an important part of the power system, transmission networks are prone to many untoward events ranging from faults, congestion, tripping, lightning strikes, etc. To provide uninterrupted and reliable service to the consumers and the producers in a power grid, an independent system operator (ISO) must be highly considerate of the transmission network security. An ample amount of hardware in the form of a protection scheme is deployed in the grid system, ensuring the health and security of transmission system and the associated equipment. The hardware comprises relays, circuit breakers, lightning arresters, etc.

This chapter examines one of the undesirable events in the transmission network known as transmission congestion. Traditionally congestion problem is solved once in every 5–30 min using tertiary dispatch. This forces the operators to operate the transmission lines conservatively much below the thermal limits to avoid congestion in between subsequent tertiary dispatches due to changes in loads, generation, and contingencies. However, real-time synchrophasor measurements of line power flows can be used to detect congestion in the smart grid and take appropriate corrective actions in real time. This will allow the lines to operate closer to the thermal limits as any violation of the limits can be resolved in real time. This chapter explains two different approaches for real-time congestion management, first a market-based scheme to encourage the electric vehicles (EVs) participate in the congestion management and then a centralized controller to resolve congestion in lesser time.

5.1 Thermal analysis of transmission lines

To understand the problem of transmission congestion, it is important to understand how the transmission lines behave in a steady-state mode of operation of the power

[1]Department of Electrical Engineering, Indian Institute of Technology Dharwad, India

grid. It is optimistic and reasonable to consider the steady-state mode of operation because the grid remains in this state for the majority of the operational time, and it is also the desirable mode of operation for the system operator. Apart from this, the modeling complexities increase as we move towards transient or other modes of operation. To reveal the behavior of the transmission lines, we consider thermal investigation of the transmission line.

A transmission line remains in the thermal stability if it follows the thermal balance equation, which states that thermal equilibrium is attained when the heat gained by a system is equal to the heat lost by the system.

The thermal gain of the transmission line is due to the copper losses i.e. I^2R losses caused by transmission line loading and ambient conditions like temperature, solar radiation, etc. not related to line loading:

$$H_{gain} = H_{load} + H_{off-load} \tag{5.1}$$

The thermal loss in a transmission line is due to radiation and convection:

$$H_{loss} = H_{rad} + H_{con} \tag{5.2}$$

For a transmission line to be in thermal equilibrium

$$H_{gain} = H_{loss} \tag{5.3}$$

$$H_{load} + H_{off-load} = H_{rad} + H_{con} \tag{5.4}$$

Considering the determining factors for all the terms in thermal equilibrium equation of the transmission line [1], (5.4) can be written and rearranged as:

$$I^2R(T_{line}) = H_{rad}(T_{line}, T_{air}) + H_{con}(T_{line}, T_{air}, \mu_{air}) - H_{off-load} \tag{5.5}$$

I is the line current; H_{con} is the thermal loss of line due to convection; T_{line} is the transmission line temperature; T_{air} is the temperature of the surrounding air; $\mu_{(air)}$ is the mobility of the air and a function of wind speed v_{wind}.

The thermal factors can be calculated as:

$$H_{off-load} = 2r \times \alpha\sigma \tag{5.6}$$

$$H_{rad} = (T_{line} - T_{air}) \left[\frac{5.278 \times 10^{-4}}{T_{avg}^{0.123}} \sqrt{\frac{Pv_{wind}}{2r}} \right] A_s \tag{5.7}$$

$$H_{con} = 5.704 \times \varepsilon \left[T_{line}'^4 - T_{air}'^4 \right] A_s \tag{5.8}$$

r is the radius of transmission line; α is the coefficient of solar radiation; σ is the radiation intensity; P is the atmospheric pressure; v_{wind} is the wind speed.

$$T_{avg} = \frac{\sum_1^n T_{line} + \sum_1^m T_{air}}{n + m}$$

A_s is the surface area of the transmission line; ε is the relative surface emissivity.

$$T'_{line} = \frac{T_{line}}{1,000} \text{Kelvin}$$

$$T'_{air} = \frac{T_{air}}{1,000} \text{Kelvin}$$

On combining (5.6), (5.7), and (5.8) with (5.5), the loading current in the transmission line will be given as:

$$I = \left[\frac{1}{R} \left((T_{line} - T_{air}) \left[\frac{5.278 \times 10^{-4}}{T_{avg}^{0.123}} \sqrt{\frac{Pv_{wind}}{2r}} \right] A_s + 5.704 \right.$$

$$\left. \times \varepsilon \left[T'_{line}{}^4 - T'_{air}{}^4 \right] A_s - 2r\alpha\sigma \right) \right]^{\frac{1}{2}} \tag{5.9}$$

In (5.9), if T_{line} is replaced with critical line temperature ($T_{line,cr}$) then the loading current I becomes the critical loading current I_{cr} of the transmission line:

$$I_{cr} = \left[\frac{1}{R} \left((T_{line,cr} - T_{air}) \left[\frac{5.278 \times 10^{-4}}{T_{avg}^{0.123}} \sqrt{\frac{Pv_{wind}}{2r}} \right] A_s + 5.704 \right.$$

$$\left. \times \varepsilon \left[T'_{line,cr}{}^4 - T'_{air}{}^4 \right] A_s - 2r\alpha\sigma \right) \right]^{\frac{1}{2}} \tag{5.10}$$

For a certain atmospheric condition, I_{cr} is the max value of current that a transmission line can handle before crossing its thermal limit.

The maximum MW rating corresponding to the critical current is:

$$P_{cr} = V_r \times I_{cr} cos(\phi_{load})$$

where V_r is the voltage at load end and ϕ_{load} is the power factor of the load.

Apart from the thermal conditions, a transmission line capability is limited by several other factors like steady-state stable limit ($p_{sssl,max}$), steady-state security limit ($P_{SL,max}$), and dynamic security limit ($P_{DSL,max}$).

The deductive arguments from Section 5.1 is that the transmission line cannot allow a power transaction which exceeds P_{cr}, $P_{sssl,max}$, and $P_{SL,max}$ or any of the them while operating in steady state and $P_{DSL,max}$ while in the dynamic state. The limiting factor $P_{line,limit}$, for the transmission line among P_{cr} and $P_{sssl,max}$, is the one which is violated first i.e.

$$P_{line,limit} = min[P_{cr}, P_{sssl,max}, P_{SL,max}, P_{DSL,max}] \tag{5.11}$$

A transmission line is said to be congested if there is a violation of the transmission line limiting factor. In congestion, a transmission line is loaded very close to or beyond the limiting factor of the line. Loading a line above its limit can cause instability in the grid due to poor damping, transient instability and also poor voltage regulation [18,26].

There may be many reasons for the violations of transmission line limiting factor like deregulation of the electricity market, increasing loads and generation on primitive networks, insertion of DERs without upgrading the transmission network, etc., but the core of this study is to identify the transmission line limit violations in the network and mitigate it, irrespective of the cause of violation. This requires continuous assessment of transmission line condition and analysis of transmission line data to check for any congestion in the network. For sensing and data collection of the transmission lines, PMUs are deployed in the network. PMUs sense and transmit data to the control center where data is processed through high-performance computational algorithms to find congested lines in the network. The controller takes the control decisions to manage the congestion in the network and relieve the transmission system of congestion.

5.2   Conventional methods for congestion control

For an efficient and reliable grid operation, it is important to keep the transmission line power within the safe limits and prevent transmission congestion. In conventional power system, several methods are used to manage the transmission congestion in power networks. These methods are broadly classified into four categories as follows:

- Network reconfiguration
- Generation rescheduling
- Load scheduling/shedding
- Hardware deployment

In this section, we will briefly review the conventional methods of congestion control and try to build the arguments for the shortcomings of these methods and build a motivation for a novel and robust congestion control technique.

Network reconfiguration

In a transmission network, there are several possible transmission routes from generation end to load end. If, for a certain power transaction between Gencom and Discom, the designated route is congested, the same transaction can be rerouted through some other network path. Alternatively, certain lines are tripped-off in a congested network, and certain lines are switched-on to change the network topology so that the modified network is not congested for a committed power transactions [7]. The former method is known as power transaction rerouting, and the latter is called transaction topology modification. Both these techniques fall under the network reconfiguration method for congestion control. Network reconfiguration is done with the consideration of the system security constraints. A genetic algorithm-based reconfiguration method evaluates the health of all possible topologies in a transmission network using synchrophasor measurements. Out of all the non-congested topologies, the one with minimum transmission loss condition is used for the power transaction [6]. A divide and conquer-based approach can be made to solve the network reconfiguration problem. Extra High Voltage (EHV) and High Voltage (HV) networks can be partitioned into zones and sub-zones.

Then optimal topology can be individually obtained for each zone and sub-zone, thereby relieving a large system from the congestion.

Generation rescheduling

Congestion management methods based on generation rescheduling cover several grid operational aspects like optimal load flow (OLF), economic dispatch (ED), market and unit commitment-based solutions, etc. The easiest way to remove congestion is to reschedule the generator power output. It is done in two prominent ways. The first way is to decrease the output of generators in one area while increasing the same amount of power in the generators of other areas. This ensures that the load is fed continuously and power balance is maintained while removing the congestion. This method is widely used for inter-area transmission congestion. In intra-area transmission congestion, this does not always work and leads to the second method. In the second method, the generator power of an area is reduced, and an equal amount of load is shut down, thereby maintaining power balance and removing congestion. In this congestion control method, care needs to be taken of the optimal power schedules of the generators. Deviation of generator power output should be done at a minimum possible economic loss. Usually, generators opt to schedule at a minimum per unit cost generations; hence an optimal rescheduling is done considering the minimal deviations of the generator from their previous schedules [8]. In the case of load tripping for an intra-area case, flexible loads are usually used, which provide flexible schedules to the grid and the consumer. Flexible loads usually consist of heating and cooling appliances. These loads can be turned on and off for a certain time without influencing the consumer's comfort. Flexibility is currently an active area of research and is beyond the scope of this chapter.

Load scheduling/shedding

Load curtailment or load shedding is the most undesired but sometimes necessary method to remove congestion from a network. In this process, both flexible and non-flexible loads are thrown off from feeding to reduce the burden on the trans-mission network. This problem is prevalent in the network is not upgraded with an increase in load and generation. In such a situation, the transmission network cannot withstand feeding the entire load at once hence throwing off the extra load on the network remains the only option. Another method is load scheduling, where loads are fed at different time intervals so that the total load on a network at a certain time is less than its transmission capability. However, this method does not preserve the needs of non-flexible loads; hence, it is used in extreme contingency cases only. Load scheduling is a prominent practice in the day-ahead market, while load shedding is a practice of the real-time market. Distribution companies are expected to submit their load schedules of the next day to the ISO before the closure of the previous day [9]. The schedules are matched with the generation schedules, and power flow is run for the network. If the network gets congested with the present schedule of loads and generations, the distribution companies may be asked to reschedule some of the loads. Again the load flow is run, and the network is checked for the congestion. Once the non-congesting schedules are reached that will be the schedules for the next day of the electricity market. This process is called load reduction bidding.

Hardware deployment

With the advent of power electronics and FACTS devices, power electronic devices have found applications in congestion management in the power transmission network. Usually, hardware-based devices improve the bus voltage profiles and enhance the power handling capability of a transmission line. The most prominent hardware solutions for transmission congestion are Interline Power Flow Controller (IPFC) [10] and Unified Power-Flow Controller (UPFC). An IPFC is voltage–source converter-based FACTS device. The device can be represented and modeled as a synchronous voltage–source injecting a controllable and near equivalent sinusoidal voltage. The voltage is controllable in both magnitude and phase angle. A controllable voltage magnitude can lead to a control on the reactive power, and a controllable phase angle gives control over the active power on a line. Based on the magnitude and phase angle of the voltage at the buses, the power handling capability of the line is improved, and thereby line power stress can be relieved on the network.

All the conventional methods of congestion control suffer a certain degree of setbacks. Network reconfiguration needs to switch on and off a series of transmission lines that may destabilize the grid [17]. Sometimes the transactions happening on the other networks may not allow the reconfiguration process. Generator power rescheduling leads to an economic loss to GENCOs due to deviation from the most economic schedules. Load scheduling leads to economic loss and raises reliability problems for Generation companies. Hardware-based solutions are limited by the limited power rating of the power electronic devices. Hardware-based solutions are also localized in control and lack global network control capability. Also, with the advancement of the conventional grid to the smart grid, conventional methods find even more glitches in implementation due to the notion of wide-area control.

The above-cited facts motivated the formulation of novel and robust congestion control methods that address most problems in conventional methods and have a global/wide-area control influence. In this chapter, we discuss two important methods of congestion control. The first method is a market-based congestion control method in which a simple but strategic energy trading is a congestion control technique. The second method is a model predictive control-based generation re-dispatch method. Both the techniques are applicable in real-time, with control over all transmission lines in a power network.

5.3 Real-time congestion control by strategic power trade with electric vehicles

Power flow in a line is sensitive to the power injected at the buses as it is a function of bus power [5]. A quantitative representation of the statement can be defined as follows:

$$(\psi_{mn})_k = \frac{\Delta P_{mn}}{\Delta P_k} \tag{5.12}$$

ψ_{mnk} is the sensitivity of the active power of line m–n to active power injection at kth bus; ΔP_{mn} is the change in the active power of line m–n; ΔP_k is the change in active power of kth bus.

The insight into (5.12) is that bus power injections can be used to control the line powers. Since the injection is the only control parameter that controls the power flow, this boils down to the fact that power trade (equivalent to bus power injection) can be a potential solution to the congestion problem. Now the elementary idea about the fixture of the problem is here, and the question remains how to execute it in the real-time power grid. The implementation process requires a control model and an execution plan.

5.3.1 Control model and PMU deployment

The control model of the system consists of two sub-models (the electric model and the market design). The two models operate together to form a joint model that can solve congestion management problems. First, we will define the models and combine them with valid arguments to make a joint model.

5.3.1.1 Modeling of congestion problem

As defined in (5.12), a power change at a bus gives a subsequent change in the line power. Since power/energy is a basic trading entity in an electricity market [20], its use for an ancillary service should be minimized. Alternately, congestion management is fortuitous on the grid rather than the desired event. Hence the power injections at the buses should be minimized as reflected in (5.13):

$$\text{Minimize} \sum_{k=1}^{n} \Delta P_k^2 \tag{5.13}$$

subject to:

$$\tilde{\psi}_I \times \Delta \tilde{P}_k = \Delta \tilde{P}_{mn}^{des} \tag{5.14}$$

$$\sum_{k=1}^{n} \Delta P_k = 0 \tag{5.15}$$

$$P_{\min} \leq \Delta P_k \leq P_{\max} \tag{5.16}$$

where

$$\tilde{\psi}_I = \begin{bmatrix} (\psi_{m_1 n_1})_1 & (\psi_{m_1 n_1})_2 & \cdots & (\psi_{m_1 n_1})_b \\ (\psi_{m_2 n_2})_1 & (\psi_{m_2 n_2})_2 & \cdots & (\psi_{m_2 n_2})_b \\ (\psi_{m_3 n_3})_1 & (\psi_{m_3 n_3})_2 & \cdots & (\psi_{m_3 n_3})_b \\ \vdots & \vdots & \cdots & \vdots \\ (\psi_{m_g n_g})_1 & (\psi_{m_g n_g})_2 & \cdots & (\psi_{m_g n_g})_b \end{bmatrix} \tag{5.17}$$

$$\Delta \widetilde{P}_k = \begin{bmatrix} \Delta P_1 \\ \Delta P_2 \\ \Delta P_3 \\ \vdots \\ \Delta P_b \end{bmatrix}, \ \Delta \widetilde{P}_{mn}^{des} = \begin{bmatrix} \Delta P_{m_1 n_1}^{des} \\ \Delta P_{m_2 n_2}^{des} \\ \Delta P_{m_3 n_3}^{des} \\ \vdots \\ \Delta P_{m_g n_g}^{des} \end{bmatrix}$$

$\widetilde{\psi}_I \in \mathbb{R}^{g \times b}$ is the generation shift factor matrix and each row of the matrix defines the sensitivity of a line with respect to bus power variation. $\Delta \widetilde{P}_k \in \mathbb{R}^{b \times 1}$ is the subsequent active power variations on buses, $i \in [1, 2, b]$. $\Delta \widetilde{P}_{mn}^{des} \in \mathbb{R}^{g \times 1}$ is the desired change in active power of each line. g is the number of congested lines and b is the total number of buses considered for congestion management.

Equation (5.14) reserves the fact that the desired power corrections are made in each congested line. Equation (5.15) preserves the power balance in the grid, such that net power injection in the grid is null and void, to prevent frequency regulation issues. Equation (5.16) preserves the security constraints of the buses.

Synchro-phasor measurements and decision making
A power system is a geographically vast and systematically bulky entity. Therefore, the sensing and protection of the system require a communication and control infrastructure. A phasor measurement unit (PMU) plays an impeccable role in wide-area sensing and monitoring. To sense and manage congestion in a bulky and widely spread power network, PMUs are deployed. PMUs sense the line power and feed the data to the control center. The control center compares the line data with the safety limits of the transmission line. If the line power data is below the safety limits, no control action is taken. However, a control decision will be taken for lines with line power exceeding the safety limits. As defined earlier, the decision will be optimal power injections at the buses. Based on the solution of the optimization problem in (5.13), the optimal injections at the buses may be negative or positive. A positive optimal injection means power injected into the grid, and a negative optimal injection means power extraction from the grid. There needs to be an injection system to inject or extract power at the buses to do the optimal injections. The injection system problem can be solved using plug-in electric vehicles capable of bidirectional trade through G2V (grid to vehicle) and V2G (vehicle to grid) connect. EVs in direct control of owners has no disposition to participate in congestion management problem of the grid. Therefore a mechanism needs to be framed that ensures the participation of the EVs in the congestion management process. This fact serves as the basis for combining a market framework with the previously defined electric model and provides a robust solution to the problem.

5.3.1.2 Market design for congestion management
Designing a market framework [15] needs the knowledge of the interests of the participants of the market [2]. Electric vehicles are expected to do the power trade if the discharging prices are high and charging prices are low at the designated injection buses at the time of congestion. Hence, the data needs to be incorporated with a time stamp that lets the controller know when and where to keep the prices low in the transmission network. To get the optimal power injections at buses and

desired power corrections in the congested lines, a strategic pricing policy is modeled as a control parameter in the market framework.

Strategic pricing policy

The strategic charging and discharging pricing functions are R_i^c and R_i^d, respectively, while R_i^{pen} serves as the strategic penalization function:

$$R_i^c = \left| \frac{\Delta P^{unmet}}{\Delta P^{desired}} \right| \times r^t + \theta sgn(\Delta P_i^{req} - \Delta P_i)(\Delta P_i^{req} - (\Delta P_i^{dis} + \Delta P_i^{ch})) \qquad (5.18)$$

$$R_i^d = \left(1 - \left| \frac{\Delta P^{unmet}}{\Delta P^{desired}} \right| \right) \times \Pi^t + \theta' exp \left[-(\Delta P_i^{req} - (\Delta P_i^{dis} + \Delta P_i^{ch}))^2 \right] \qquad (5.19)$$

$$R_i^{pen} = r^t + \theta' |\Delta P_i^{req}| \qquad (5.20)$$

$r^t, \Pi^t, \theta, \theta'$ are the per unit energy pricing parameters (₹/unit) at time "t". ΔP_i^{dis} and ΔP_i^{ch} are the EV discharging and charging power at time "t". The discharging power is always considered to be positive while the charging power is considered to be negative. ΔP_i^{req} is defined as the power correction requested by the grid to the EVs at the ith bus and $\Delta P_i = \Delta P_i^{dis} + \Delta P_i^{ch}$:

$$Sgn(x) = \left\{ \begin{array}{l} 1, x \geq 0 \\ -1, x < 0 \end{array} \right., exp\,(-x^2) = \left\{ \begin{array}{l} 1, x = 0 \\ < 1, x \neq 0 \end{array} \right.$$

To get a detailed insight of the pricing policy and the market framework we do an analysis of the pricing functions.

Both the charging and discharging prices are designed as strategic rewards while the penalization function is designed as a strategic penalty. The reward functions comprise two parts as defined below:

- Individual gain factor $\Rightarrow \theta sgn(\Delta P_i^{req} - \Delta P_i)(\Delta P_i^{req} - (\Delta P_i^{dis} + \Delta P_i^{ch}))$
- Overall gain factor $\Rightarrow \left| \frac{\Delta P^{unmet}}{\Delta P^{desired}} \right| \times r^t$

Considering the strategic charging pricing function at ith bus defined in (5.18), if the EVs are able to provide an exact amount of power that is requested at the bus i.e. $\Delta P_i^{dis} + \Delta P_i^{ch} = \Delta P_i^{req}$, then

$$\theta sgn(\Delta P_i^{req} - \Delta P_i)(\Delta P_i^{req} - (\Delta P_i^{dis} + \Delta P_i^{ch})) = 0 \qquad (5.21)$$

This means that the EVs are able to minimize the individual gain factor.

If the corrections made in the line powers by bus injections be defined as

$$\Delta \widetilde{P}_{mn}^{met} = \begin{bmatrix} \Delta P_{m_1 n_1}^{met} \\ \Delta P_{m_2 n_2}^{met} \\ \Delta P_{m_3 n_3}^{met} \\ \vdots \\ \Delta P_{m_g n_g}^{met} \end{bmatrix} \qquad (5.22)$$

Then the unfulfilled line power corrections are defined as;

$$
\Delta \widetilde{P}_{mn}^{unmet} =
\begin{bmatrix}
\Delta P_{m_1 n_1}^{des} - \Delta P_{m_1 n_1}^{met} \\
\Delta P_{m_2 n_2}^{des} - \Delta P_{m_2 n_2}^{met} \\
\Delta P_{m_3 n_3}^{des} - \Delta P_{m_3 n_3}^{met} \\
\vdots \\
\Delta P_{m_g n_g}^{des} - \Delta P_{m_g n_g}^{met}
\end{bmatrix}
=
\begin{bmatrix}
\Delta P_{m_1 n_1}^{unmet} \\
\Delta P_{m_2 n_2}^{unmet} \\
\Delta P_{m_3 n_3}^{unmet} \\
\vdots \\
\Delta P_{m_g n_g}^{unmet}
\end{bmatrix}
\tag{5.23}
$$

Defining

$$
\Delta P^{desired} = \|\widetilde{\Delta P}_{mn}^{des}\|_2 = \sqrt{(\Delta P_{m_1 n_1}^{des})^2 + (\Delta P_{m_2 n_2}^{des})^2 \ldots + (\Delta P_{m_g n_g}^{des})^2} \tag{5.24}
$$

$$
\Delta P^{unmet} = \|\Delta \widetilde{P}_{mn}^{unmet}\|_2 \tag{5.25}
$$

If all the buses are able to fulfill the requested power, then

$$
\widetilde{\Psi}_I \times \widetilde{\Delta P}_k = \widetilde{\Delta P}_{mn}^{des} = \widetilde{\Delta P}_{mn}^{met} \tag{5.26}
$$

$$
\Rightarrow
$$

$$
\Delta \widetilde{P}_{mn}^{unmet} =
\begin{bmatrix}
0 \\
0 \\
0 \\
\vdots \\
0
\end{bmatrix}
\tag{5.27}
$$

which means

$$
\left| \frac{\Delta P^{unmet}}{\Delta P^{desired}} \right| \times r^t = 0 \tag{5.28}
$$

In such a case $R_i^c = 0$, which means on complete congestion management, the EVs will be charging at free of cost. For the ease of understanding a plot of R_i^c against ΔP_i is shown in Figure 5.1, where $\Delta P_i^{req} = -0.4931$ p.u. It is observed that the charging price is minimum when the EVs are exactly fulfilling the grid request ΔP_i^{req} at the bus.

Similarly the strategic discharging pricing function can be analyzed as a two-part function defined as

- Individual gain factor $\Rightarrow \theta' exp[-(\Delta P_i^{req} - (\Delta P_i^{dis} + \Delta P_i^{ch}))^2]$
- Overall gain factor $\Rightarrow \left(1 - |\frac{\Delta P^{unmet}}{\Delta P^{desired}}|\right) \times \Pi^t$

If $(\Delta P_i^{dis} + \Delta P_i^{ch})^2 = \Delta P_i^{req}$

$$
\Rightarrow \theta' exp\left[-(\Delta P_i^{req} - (\Delta P_i^{dis} + \Delta P_i^{ch}))^2\right] = \theta' exp(0) = \theta' \tag{5.29}
$$

This maximizes the individual gain factor of the discharging price.

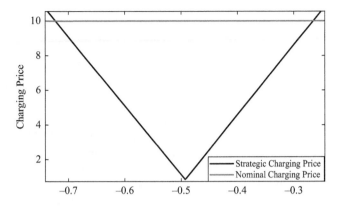

Figure 5.1   Strategic charging price R_i^c as a function of bus power injection

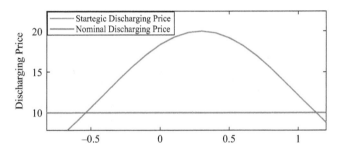

Figure 5.2   Strategic discharging price R_i^d as a function of bus power injections

Considering (5.26), $\left|\frac{\Delta P^{unmet}}{\Delta P^{desired}}\right| = 0$

$$\Rightarrow \left(1 - \left|\frac{\Delta P^{unmet}}{\Delta P^{desired}}\right|\right) \times \Pi^t = \Pi^t \qquad (5.30)$$

This maximizes the overall gain factor of the discharging price. In this case, the discharging price will be $R_i^d = \Pi^t + \theta'$. A plot of strategic discharging price against the bus power corrections for a requested bus injection of 0.2943 p.u. is shown in Figure 5.2. It is observed that the maximum value of the discharging price is when the bus fulfills the grid request properly.

5.3.1.3   Joint model through game formulation

The sub-models defined in previously have to operate together as a joint model for a robust solution of the problem. To integrate the two sub-models into a joint model, game theoretic approach is used [16,19].

An ordinal game frame G_i at ith bus is a space defined on ternate parameters as follows:

$$G_i = \langle N_i, \widehat{S}_{n,i}, O_{n,i} \rangle \qquad (5.31)$$

N_i is the number of players (EVs) in the game at ith bus; $\widehat{S}_{n,i}$ is the strategy set of the nth players at ith bus i.e. charging ($c_{n,i}^t > 0$), discharging ($c_{n,i}^t < 0$) and layoff ($c_{n,i}^t = 0$), $O_{n,i}$ is the payoff of nth EV at ith bus.

Based on the optimal bus injections, buses may get either $\Delta P_i^* > 0$ (discharging) or $\Delta P_i^* < 0$ (charging) as a power correction request. This leads to allotment of expected strategies, to the busses and are represented as S_i^{exp} defining the expectation of the grid from the EVs at the buses.

To gain control over EVs by the grid, payoff settlement is done to give the EVs incentive or penalize based on the strategies of EVs and the allotted expected strategies at the buses (Table 5.1). Provided the expected strategy at the bus, if the strategy of EVs match with it, the EVs will be incentivized and if the EV strategy is opposite to the expected strategy, the EV will be penalized. For a layoff condition of the EV, there is neither incentive nor penalty (Table 5.2).

Execution of joint model
The market model for transmission congestion management problem is formulated based on the greedy behavior of the EV owners. Since EVs are in direct control of the EV owners, we consider the owner's greed as the EV greed for technical ease. Vehicle to grid (V2G) majorly has two interactions with the power grid, i.e., charging and discharging; considering the greedy behavior of the EVs, they wish to either charge free of cost or discharge at higher prices. So, to fulfill their greed, they

Table 5.1   Expected strategy distribution

S. no.	Optimized bus solution (ΔP_i^*)	Allotted expected strategy (S_i^{exp})
1	$\Delta P_i^* > 0$	$c_{n,i}^t < 0$ (Discharging)
2	$\Delta P_i^* < 0$	$c_{n,i}^t > 0$ (Charging)
3	$\Delta P_i^* = 0$	No allocation

Table 5.2   Strategic payoff settlement

S. no.	Allotted expected strategy(S_i^{exp})	EV strategy $S_{n,i}^{ev}$	Payoff $O_{n,i}$
1	$c_{n,i}^t > 0$	$c_{n,i}^t > 0$	R_i^c
2		0	0
3		$c_{n,i}^t < 0$	R_i^{pen}
4	$c_{n,i}^t < 0$	$c_{n,i}^t > 0$	R_i^{pen}
5		0	0
6		$c_{n,i}^t < 0$	R_i^d

will try to adopt all the strategies that follow one of the two conditions stated below:

$$S_n^*(S_i^{exp}, S_{-n}) = arg \max_{\widehat{S}_{n,i}} (R_i^d) \tag{5.32}$$

$$S_n^*(S_i^{exp}, S_{-n}) = arg \min_{\widehat{S}_{n,i}} (R_i^c) \tag{5.33}$$

S_n^* is the best strategy of nth EV in response to an allotted Expected strategy S_i^{exp} and the strategy adopted by all other EVs except for nth EV S_{-n} at the ith bus. To fulfill their greed and achieve more significant incentives, the EVs will try to adopt the best strategic response.

Equation (5.32) corresponds to the statement that the best strategy of a charging EV will be to minimize its charging price while 5.33 defines the fact that for a discharging vehicle, the best strategy is to maximize its discharging price. To understand the set of strategies that correspond to the fulfillment of interest of the EVs, we define Nash equilibrium.

Nash equilibrium: The Nash equilibrium of a game is defined as the set of strategies from which a unilateral deviation of a player in the game does not result in an additional incentive for the player.

For a successful execution of the joint model, we need to prove three fundamentally important things as defined below.

1. EVs are benefited by the joint model
2. The EV to EV game reaches a Nash equilibrium
3. Nash equilibrium is a solution to the congestion management problem

We define Nash equilibrium $(\Delta P_i^{dis} + \Delta P_i^{ch})^*$ of EV to EV game that is as follows:
When $\Delta P_i^{avl} = \Delta P_i^{req}$.

$$(\Delta P_i^{dis} + \Delta P_i^{ch})^* = \Delta P_i^{req} \tag{5.34}$$

When $\Delta P_i^{avl} \neq \Delta P_i^{req}$.

$$(\Delta P_i^{dis} + \Delta P_i^{ch})^* = \min[|\Delta P_i^{req}|, |\Delta P_i^{avl}|] \tag{5.35}$$

where ΔP_i^{avl} is the total power correction ability of EVs at ith bus.

Equations (5.34) and (5.35) define the Nash equilibrium for the proposed game pertaining to two possible cases of the game.

5.3.2 Proof

We will try to validate the fact that:

- Nash equilibrium maximizes the EV incentive.
- Nash equilibrium is a solution to the congestion problem.

If $\Delta P_i^{dis} + \Delta P_i^{ch} = \Delta P_i^{avl} = \Delta P_i^{req}$ is valid for all buses, then for all the buses:

$$(\Delta P_i^{req} - \Delta P_i)sgn(\Delta P_i^{req} - (\Delta P_i^{dis} + \Delta P_i^{ch})) = 0 \tag{5.36}$$

And

$$exp\left[-(\Delta P_i^{req} - (\Delta P_i^{dis} + \Delta P_i^{ch}))^2\right] = 1 \tag{5.37}$$

Equations (5.36) and (5.37) bear the form $xsgn(x)$ and e^{-x^2} we know that

$$\min_{x} xsgn(x) = 0$$

$$\max_{x} e^{-x^2} = 1$$

As $\Delta P_i^{dis} + \Delta P_i^{ch} = \Delta P_i^{req}$ is true for all buses then $\widetilde{\Phi}_I \times \Delta \widetilde{P}_k = \Delta \widetilde{P}_{mn}^{des} = \Delta \widetilde{P}_{mn}^{met}$. As $\Delta \widetilde{P}_{mn}^{des} = \Delta \widetilde{P}_{mn}^{met}$, then $\Delta \widetilde{P}_{mn}^{unmet}$ is a null vector and $\Delta P^{unmet} = 0$ which means

$$\left|\frac{\Delta P^{unmet}}{\Delta P^{desired}}\right| = 0 \ and \left(1 - \left|\frac{\Delta P^{unmet}}{\Delta P^{desired}}\right|\right) = 1$$

considering the design of R_i^c and R_i^d, the strategy $\Delta P_i^{dis} + \Delta P_i^{ch} = \Delta P_i^{req}$ is the Nash equilibrium of the EV to EV game that provides a solution to congestion problem.

Considering the second case, $\Delta P_i^{avl} \neq \Delta P_i^{req}$. The possibilities of $\Delta P_i^{avl} \neq \Delta P_i^{req}$ are either $\Delta P_i^{avl} < \Delta P_i^{req}$ or $\Delta P_i^{avl} > \Delta P_i^{req}$. This means that in such a case, there is a chance that $\Delta P^{unmet} \neq 0$ though $\Delta P^{unmet} < \Delta P^{desired}$ as EVs do some correction, pertaining to the fact

$$0 < \left|\frac{\Delta P^{unmet}}{\Delta P^{desired}}\right| < 1 \tag{5.38}$$

$$0 < \left|1 - \frac{\Delta P^{unmet}}{\Delta P^{desired}}\right| < 1 \tag{5.39}$$

As deviation from ΔP_i^{req} increases, the incentive decreases as seen from Figures 5.1 and 5.2. Also, remaining closer of EVs to the requested value ΔP_i^{req}, the congestion clearance happens properly. The deviation from ΔP_i^{req} is minimized by adopting (5.40):

$$(\Delta P_i^{dis} + \Delta P_i^{ch})^* = \min\left[|\Delta P_i^{req}|, |\Delta P_i^{avl}|\right] \tag{5.40}$$

Hence (5.40) becomes the Nash equilibrium for Case 2.

For the ease of understanding, a flowchart is presented in Figure 5.3 to realize the process flow of proposed algorithm.

5.3.3   Test results

The algorithm is validated by testing on IEEE-16 Machine 68-Bus testbed [8]. The transmission lines between the buses 27-53 and 53-54 were considered to be congested in Figure 5.4. Buses 17, 2–28, 33, 36, 40, 47, 48, 53, 56, 59–61,

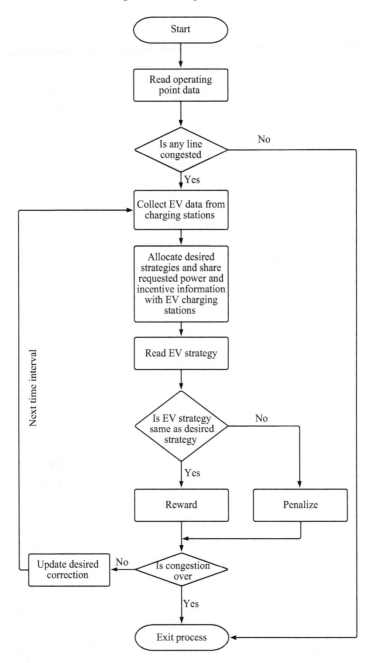

Figure 5.3   Algorithm for proposed congestion management method

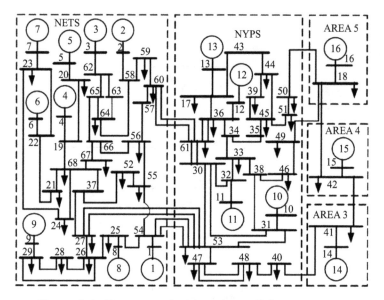

Figure 5.4   Test system-2: 16-machines 68-bus test system

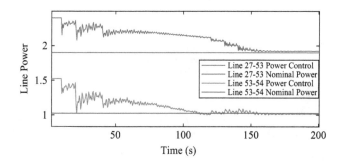

Figure 5.5   Multiple transmission line congestion management

and 64 are considered to have V2G interfaces. The allotted expected strategy, buses 25, 26, 27, 28, and 64 is charging, hence EVs are expected to charge, while at the rest of the buses, the EVs are expected to discharge. Charging price at bus 25 and discharging price at bus 60 is tracked. Participation of EVs at different buses is considered at different time intervals in a time frame of 200 s in the congestion management game. Nominal operating points of congested lines is at 1.89 p.u. and 1.102 p.u. respectively, and the desired congestion clearance of −0.5 p.u. each. As observable from Figure 5.5, congestion in both the lines is successfully managed in both the lines at the same time. Figures 5.6 and 5.7 provide the evidence of EVs getting incentive by charging at minimized and discharging at maximized prices. The proposed algorithm was also validated for the influence of uncertainty in the power output of renewable energy resources. Nine DFIGs were installed at buses 21, 28, 33, 37, 40, 42, 52, and 56 in the testbed and considered under variable wind speeds. The influence of variable power output of the DFIGs was studied on lines 60-61 and the proposed method was used to resolve the issue. Proposed algorithm is repeated after interval of 1-min time frame and 30-s time frame and the results are depicted in Figure 5.8. The analysis for RER fluctuations and control is given in Table 5.3.

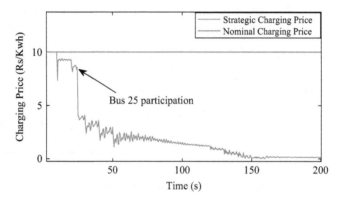

Figure 5.6   Dynamics of charging price at bus 25

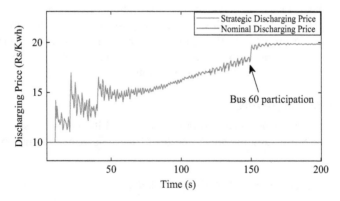

Figure 5.7   Dynamics of discharging price at bus 60

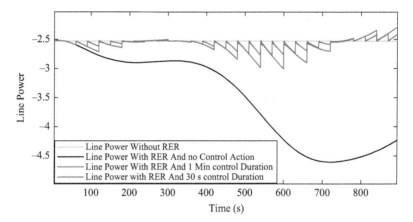

Figure 5.8   Tie-line power flow with variable renewable power output

Table 5.3   *Analysis for congestion control under DFIG power fluctuations*

S. no.	Parameter	No control action	1 Min control action	30 s control action
1	Line power flow deviation (RMS)	1.2288	0.1420	0.0723
2	Max line flow deviation	−2.0787	−0.4788	−0.2466

5.4   Model predictive control-based generation re-dispatch for real-time congestion management

This section explains the modeling aspects of the power system for congestion management and the design of a model predictive controller (MPC) [3,4] to regulate the line power flows. The state-space model establishes dynamic relations between resource inputs, i.e., synchronous machine (SM) and battery energy storage systems (BESS) to power flow in the transmission lines of the power system. The MPC is able to generate the optimal sequence of the control signals while considering the constraints of SMs and BESSs, i.e., maximum and minimum magnitude constraints as well as rate constraints.

5.4.1   System modeling

The dynamic relation between line power flows and the control resource inputs, including parametric variation and other disturbances, is expressed mathematically as

$$\dot{\mathbf{x}}_c(t) = \mathbf{A}_c \mathbf{x}_c(t) + \mathbf{B}_c \mathbf{u}_c(t) + \mathbf{B}_{c1} \mathbf{u}_{c1}(t) + \mathbf{d}_c(t) \tag{5.41}$$

$$\mathbf{y}_c(t) = \mathbf{C}_c \mathbf{x}_c(t) + \mathbf{D}_{c1} \mathbf{u}_{c1}(t) \tag{5.42}$$

where $\mathbf{A}_c$ is a system state matrix of order $\mathbb{R}^{n_t \times n_t}$, n_t is total states of the system, $\mathbf{B}_c$ is a system input matrix with order of $\mathbb{R}^{n_t \times m}$, p is total number of inputs of the system, $\mathbf{C}_c$ is output matrix of order $\mathbb{R}^{q \times n_t}$, total number of outputs are q. $\mathbf{d}_c(t) = \Delta A_c \mathbf{x}_c(t) + \mathbf{d}(t)$ is the disturbance matrix including the parametric variation in the system and other disturbances and it is bounded in nature. $\mathbf{B}_{c1}$ and $\mathbf{D}_{c1}$ are disturbance matrices and bounded in nature, and have order $\mathbb{R}^{n_t \times n_d}$, $\mathbb{R}^{q \times n_d}$ respectively, and n_b is total disturbances inputs. In this context total states n_t includes all the states of SMs, i.e., n_g, and BESS , i.e., n_b. The total number of outputs, i.e., q, equal to the power flows in the lines considered. The inputs, i.e., p, includes SMs and/or BESS inputs.

To perform real-time congestion management in the power system, the power system operator sends a set of optimized control outputs generated by MPC in the form of a reference vector $\mathbf{r}(t)$ to the selected SMs/BESS. The selected resources keep track of power flows in lines as per the reference and help manage the power without reaching abnormal values that lead to system failure. The MPC requires an accurate, detailed model of the system. The subsections below elaborate on developing the detailed power system required for line flow control to manage congestion. The MPC mathematical descriptions are also elaborated in the following sub-sections.

5.4.1.1   State space model of line power dynamics

The models of the power system elements such as SMs and BESS are included in the Appendices 5.1–5.3. Every vector and matrix of the system that establishes the connection between resources (i.e., SMs and/or BESS) inputs to lines power flow outputs w.r.t (5.41) and (5.42) is elaborately written as:

All states of the system are included in state vector $\mathbf{x}_c(t) = [\mathbf{x}_1(t) \ldots \mathbf{x}_i(t) \ldots \mathbf{x}_g(t) | \mathbf{x}_b(t)]^T$. Where all BESS-related states are present in $\mathbf{x}_b(t) = [\Delta P_{B_1}(t) \ldots \Delta P_{B_{n_b}}(t)]^T$, $\Delta P_{B_{n_b}}(t)$ power output of n_b^{th} BESS, all the states corresponding to i^{th} SM are represented in $\mathbf{x}_1(t) = [\Delta \omega_i(t), \Delta \delta_i(t), \Delta P_{SVi}(t), \Delta P_{CHi}(t), \Delta P_{Mi}(t)]^T$, $\Delta \omega_i(t)$ speed deviation in p.u., $\Delta \delta_i(t)$ change in rotor angle in p.u., $\Delta P_{SVi}(t)$ governors valve position in p.u., $\Delta P_{CHi}(t)$ turbine steam chest output in p.u., $\Delta P_{Mi}(t)$ mechanical power in p.u. All the control inputs are included in the vector $\mathbf{u}_c(t) = [u_1(t) \ldots u_i(t) \ldots u_g(t) | \mathbf{u}_b(t)]^T \in \mathbb{R}^m$, where $u_i(t) = \Delta P_{Ci}(t)$, is the load reference coming from system operator for ith SM, $\mathbf{u}_b(t) = [\Delta P_{B_1}^*(t) \ldots \Delta P_{B_i}^*(t) \ldots \Delta P_{B_{n_b}}^*(t)] \in \mathbb{R}^{n_b}$, $\Delta P_{B_i}^*(t)$ reference given to ith BESS. All the disturbance inputs, in this context DFIG power injections, to the system at various buses are included in vector $\mathbf{u}_{c1} = [\Delta P_1, \ldots, \Delta P_{n_d}]$, ΔP_{n_d} is the change in n_d^{th} DFIG disturbance input. All the line power flows to be controlled in the system are included in the vector $\mathbf{y}_c(t) = [\Delta P_{m_1 n_1}, , \Delta P_{m_i n_i}, , \Delta P_{m_q n_q}]^T$. Where $\Delta P_{m_i n_i}$ is the deviation in the power flow through line between the m_i^{th} and n_i^{th} buses.

The state matrix and the corresponding entries are as follows:

$$\mathbf{A}_c = \begin{bmatrix} \mathbf{A}_{GG} & \mathbf{A}_{GB} \\ \mathbf{O} & \mathbf{A}_{BB} \end{bmatrix}, \text{where,} \qquad \mathbf{A}_{GG} = \begin{bmatrix} \mathbf{A}_{11} & \cdots & \mathbf{A}_{1i} & \cdots & \cdots & \mathbf{A}_{1g} \\ & \ddots & \vdots & \vdots & \vdots & \vdots \\ \mathbf{A}_{i1} & & \mathbf{A}_{ii} & \cdots & \mathbf{A}_{ij} & \mathbf{A}_{ig} \\ \vdots & & \vdots & & \vdots & \vdots \\ \mathbf{A}_{g1} & \cdots & \mathbf{A}_{gi} & \cdots & \cdots & \mathbf{A}_{gg} \end{bmatrix},$$

$$\mathbf{A}_{ii} = \begin{bmatrix} -\dfrac{D_i}{2H_i} & -\dfrac{\omega_B}{2H_i}\left(\dfrac{\partial P_{ei}}{\partial \delta_i}\right) & 0 & \dfrac{1}{2H_i}K_{HP_i} & \dfrac{1}{2H_i}(1-K_{HP_i}) \\ \omega_B & 0 & 0 & 0 & 0 \\ \dfrac{-1}{R_{D_i}T_{SV_i}} & 0 & \dfrac{-1}{T_{SV_i}} & 0 & 0 \\ 0 & 0 & \dfrac{1}{T_{CH_i}} & \dfrac{-1}{T_{CH_i}} & 0 \\ 0 & 0 & 0 & \dfrac{1-K_{HP_i}}{T_{RH_i}} & \dfrac{-1}{T_{RH_i}} \end{bmatrix},$$

$\frac{\partial P_{ei}}{\partial \delta_i}$ is the ith SM output sensitivity w.r.t its rotor angle, T_{SV_i} is the ith SM governor time constant in second, P_{ei} is the ith SM output in p.u., ω_B is the base speed of all SMs expressed in rad/s,

$$\mathbf{A}_{ij} = \begin{bmatrix} 0 & -\dfrac{\omega_B}{2H_i}\left(\dfrac{\partial P_{ei}}{\partial \delta_j}\right) & 0 & 0 & 0 \\ 0 & 0 & 0 & 0 & 0 \\ 0 & 0 & 0 & 0 & 0 \\ 0 & 0 & 0 & 0 & 0 \\ 0 & 0 & 0 & 0 & 0 \end{bmatrix},$$

$\frac{\partial P_{ei}}{\partial \delta_j}$ is the ith SM output sensitivity w.r.t. jth SM rotor angle,

$$\mathbf{A}_{GB} = \begin{bmatrix} \mathbf{A}_{1b} \\ \vdots \\ \mathbf{A}_{ib} \\ \vdots \\ \mathbf{A}_{gb} \end{bmatrix} \in \mathbb{R}^{n_g \times n_b},$$

$$\mathbf{A}_{ib} = \begin{bmatrix} -\dfrac{\omega_B}{2H_i}\left(\dfrac{\partial P_{ei}}{\partial P_{B_1}}\right) & -\dfrac{\omega_B}{2H_i}\left(\dfrac{\partial P_{ei}}{\partial P_{B_{n_b}}}\right) \\ 0 & 0 \\ 0 & 0 \\ 0 & 0 \\ 0 & 0 \end{bmatrix} \in \mathbb{R}^{5 \times n_b},$$

$\frac{\partial P_{ei}}{\partial P_{B_{n_b}}}$ is the ith SM output sensitivity w.r.t. n_b^{th} BESS output power,

$$\mathbf{A}_{BB} = \begin{bmatrix} -\dfrac{1}{\tau_1} & & \\ & \ddots & \\ & & -\dfrac{1}{\tau_{n_b}} \end{bmatrix},$$

τ_{n_b} is the n_b^{th} BESS time constant in seconds. The input matrix and the corresponding entries are as follows:

$$\mathbf{B}_c = \begin{bmatrix} \mathbf{B}_{GG} & \mathbf{O} \\ \mathbf{O} & \mathbf{B}_{BB} \end{bmatrix},$$

Where

$$\mathbf{B}_{GG} = \begin{bmatrix} \mathbf{B}_1 \\ \vdots \\ \mathbf{B}_i \\ \vdots \\ \mathbf{B}_g \end{bmatrix} \in \mathbb{R}^{n_t \times m}, \mathbf{B}_{BB} = \begin{bmatrix} \dfrac{1}{\tau_1} & & \\ & \ddots & \\ & & \dfrac{1}{\tau_{n_b}} \end{bmatrix} \in \mathbb{R}^{n_b \times n_b},$$

$$\mathbf{B}_i = \begin{bmatrix} 0 & \cdots & 0 & \cdots & 0 \\ 0 & \cdots & 0 & \cdots & 0 \\ 0 & \cdots & (1/T_{SV_i})_{(3,i)} & \cdots & 0 \\ 0 & \cdots & 0 & \cdots & 0 \\ 0 & \cdots & 0 & \cdots & 0 \end{bmatrix} \in \mathbb{R}^{5 \times m},$$

T_{SV_i} is the ith SM governor time constant in sec.

The output matrix and the corresponding entries are as follows:

$$\mathbf{C}_c = \begin{bmatrix} \mathbf{C}_{GG} \\ \mathbf{C}_{BB} \end{bmatrix}^T,$$

where

$$\mathbf{C}_{GG} = \begin{bmatrix} \mathbf{C}_1 \\ \vdots \\ \mathbf{C}_i \\ \vdots \\ \mathbf{C}_g \end{bmatrix}^T \in \mathbb{R}^{q \times n_g}, \mathbf{C}_i = \begin{bmatrix} 0 & \left(\dfrac{\partial P_{(m_1,n_1)}}{\partial \delta_i}\right) & 0 & 0 & 0 \\ \vdots & \vdots & \vdots & \vdots & \vdots \\ 0 & \left(\dfrac{\partial P_{(m_q,n_q)}}{\partial \delta_i}\right) & 0 & 0 & 0 \end{bmatrix} \in \mathbb{R}^{q \times 5},$$

$\frac{\partial P_{(m_q,n_q)}}{\partial \delta_i}$ is the power flow sensitivity of line between the buses m_q and n_q w.r.t. ith SM rotor angle.

$$\mathbf{C}_{BB} = \begin{bmatrix} \left(\dfrac{\partial P_{(m_1,n_1)}}{\partial P_{B_1}} \right) & \cdots & \left(\dfrac{\partial P_{(m_1,n_1)}}{\partial P_{B_{n_b}}} \right) \\ \vdots & \vdots & \vdots \\ \left(\dfrac{\partial P_{(m_q,n_q)}}{\partial P_{B_1}} \right) & \cdots & \left(\dfrac{\partial P_{(m_q,n_q)}}{\partial P_{B_{n_b}}} \right) \end{bmatrix} \in \mathbb{R}^{q \times n_b},$$

$\frac{\partial P_{(m_q,n_q)}}{\partial P_{n_d}}$ is the power flow sensitivity of line between the buses m_q and n_q w.r.t. n_d^{th} DFIG power injection. The input disturbance matrix and corresponding entries are as follows:

$$\mathbf{B}_{c1} = \begin{bmatrix} \mathbf{B}_{11,} \\ \vdots \\ \mathbf{B}_{i1} \\ \vdots \\ \mathbf{B}_{g1} \end{bmatrix} \in \mathbb{R}^{n_t \times n_b},$$

where

$$\mathbf{B}_{i1} = \begin{bmatrix} -\dfrac{\omega_B}{2H_i} \left(\dfrac{\partial P_{ei}}{\partial P_1} \right) & \cdots & -\dfrac{\omega_B}{2H_i} \left(\dfrac{\partial P_{ei}}{\partial P_{n_d}} \right) \\ 0 & \cdots & 0 \\ 0 & \cdots & 0 \\ 0 & \cdots & 0 \\ 0 & \cdots & 0 \end{bmatrix} \in \mathbb{R}^{5 \times n_d},$$

$\frac{\partial P_{ei}}{\partial P_{n_d}}$ is the ith SM output sensitivity w.r.t. n_d^{th} DFIG disturbance power injection. The output disturbance matrix and its entries are as follows:

$$\mathbf{D}_{c1} = \begin{bmatrix} \left(\dfrac{\partial P_{(m_1,n_1)}}{\partial P_1} \right) & \cdots & \left(\dfrac{\partial P_{(m_1,n_1)}}{\partial P_{n_d}} \right) \\ \vdots & \vdots & \vdots \\ \left(\dfrac{\partial P_{(m_q,n_q)}}{\partial P_1} \right) & \cdots & \left(\dfrac{\partial P_{(m_q,n_q)}}{\partial P_{n_d}} \right) \end{bmatrix} \in \mathbb{R}^{q \times n_d}$$

where $\frac{\partial P_{(m_q,n_q)}}{\partial P_{n_d}}$ is the power flow sensitivity of line between buses m_q and n_q w.r.t. DFIG disturbance power injection. The MPC takes a detailed system model and produces a sequence of control signals. For the proper functioning of the controller, the system must satisfy stabilizability and detectability criteria.

Mathematical formulation of line power flow and SM output sensitivities

The overall power system has $\mathfrak{B}$ buses, g SMs, and n_b BESS. The current injections and voltage magnitudes at various buses related as

$$\mathbf{I}_{\mathfrak{B}} = \mathbf{Y} * \mathbf{V}_{\mathfrak{B}} \tag{5.43}$$

$$\mathbf{V}_{\mathfrak{B}} = \mathbf{Z} * \mathbf{I}_{\mathfrak{B}} \tag{5.44}$$

m^{th} bus voltage is

$$
\begin{aligned}
V_m &= \sum_{k=1}^{g} E_k^0 |Z_{m\mathfrak{B}_k}| |y_k| e^{j\left(\delta_k + \delta_k^0 - \frac{\pi}{2} + \alpha_k + \theta_{m\mathfrak{B}_k}\right)} \\
&\quad + \sum_{k_b=1}^{n_b} \frac{P_{B_{k_b}}}{|V_{k_b}|} |Z_{mk_b}| e^{j(\theta_{mk_b} + \theta_{k_b})} \\
&= \sum_{k=1}^{g} |V_{m\mathfrak{B}_k}| e^{j\left(\delta_k + \delta_k^0 - \frac{\pi}{2} + \alpha_k + \theta_{m\mathfrak{B}_k}\right)} \\
&\quad + \sum_{k_b=1}^{n_b} \frac{P_{B_{k_b}}}{|V_{k_b}|} |Z_{mk_b}| e^{j(\theta_{mk_b} + \theta_{k_b})} \\
&= V_{m,real} + jV_{m,imag}
\end{aligned} \tag{5.45}
$$

where $\mathfrak{B}_k$ is the kth SM bus, $|Z_{m\mathfrak{B}_k}|$ is the magnitude of $(m, \mathfrak{B}_k)^{th}$ element of Z matrix, $\theta_{m\mathfrak{B}_k}$ is the phase angle of the same element, $|V_{k_b}|$ is the voltage present at k_b^{th} bus of BESS, θ_{k_b} is the phase angle of the same, $|y_k|$ is the kth SM admittance, α_k is the angle of the same, and $V_{m,real}$, $V_{m,imag}$ are expressed as follows:

$$
\begin{aligned}
V_{m,real} &= \sum_{k=1}^{g} |V_{m\mathfrak{B}_k}| \cos\left(\delta_k + \delta_k^0 - \frac{\pi}{2} + \alpha_k + \theta_{m\mathfrak{B}_k}\right) \\
&\quad + \sum_{k_b=1}^{n_b} \frac{P_{B_{k_b}}}{|V_{k_b}|} \cos(\theta_{mk_b} + \theta_{k_b})
\end{aligned} \tag{5.46}
$$

$$
\begin{aligned}
V_{m,imag} &= \sum_{k=1}^{g} |V_{m\mathfrak{B}_k}| \sin\left(\delta_k + \delta_k^0 - \frac{\pi}{2} + \alpha_k + \theta_{m\mathfrak{B}_k}\right) \\
&\quad + \sum_{k_b=1}^{n_b} \frac{P_{B_{k_b}}}{|V_{k_b}|} \sin(\theta_{mk_b} + \theta_{k_b})
\end{aligned} \tag{5.47}
$$

The real-power flow between m^{th} and n^{th} bus is

$$P_{mn} = Re\left\{ V_m \left\{ [V_m - V_n] y_{mn} + jV_m \frac{B_{mn}}{2} \right\}^* \right\} \tag{5.48}$$

where $y_{mn} = G_{mn} + jB_{mn}$ is the admittance, V_n is the voltage at nth bus and its expression is obtained by modifying (5.46). The power flow equation in (5.48) is further simplified using (5.46) and (5.47) as follows:

$$P_{mn} = V_m^2 G_{mn} - V_{m,real}V_{n,real}G_{mn} - V_{m,imag}V_{n,imag}G_{mn}$$
$$+ V_{m,real}V_{n,imag}B_{mn} - V_{n,real}V_{m,imag}B_{mn} \tag{5.49}$$

The constant part of the angle in bus voltage equations in (5.46) and (5.47) is defined as $\beta_{m\mathcal{B}_k} = \delta_k^0 - \frac{\pi}{2} + \alpha_k + \theta_{m\mathcal{B}_k}$.

The sensitivity of real power flow in (5.49) w.r.t. the ith SM load angle is expressed as follows:

$$\frac{\partial P_{mn}}{\partial \delta_i} = G_{mn}|V_{n\mathcal{B}_i}|V_{m,real}\sin\left(\beta_{n\mathcal{B}_i} + \delta_i\right) + G_{mn}|V_{m\mathcal{B}_i}|V_{n,real}\sin\left(\beta_{m\mathcal{B}_i} + \delta_i\right) -$$

$$G_{mn}|V_{n\mathcal{B}_i}|V_{n,imag}\cos\left(\beta_{ni} + \delta_i\right) - G_{mn}|V_{m\mathcal{B}_i}|V_{n,imag}\cos\left(\beta_{m\mathcal{B}_i} + \delta_i\right) +$$

$$B_{mn}|V_{n\mathcal{B}_i}|V_{m,real}\cos\left(\beta_{n\mathcal{B}_i} + \delta_i\right) - B_{mn}|V_{m\mathcal{B}_i}|V_{n,imag}\sin\left(\beta_{m\mathcal{B}_i} + \delta_i\right) -$$

$$B_{mn}|V_{m\mathcal{B}_i}|V_{n,real}\cos\left(\beta_{m\mathcal{B}_i} + \delta_i\right) + B_{mn}|V_{n\mathcal{B}_i}|V_{m,imag}\sin\left(\beta_{n\mathcal{B}_i} + \delta_i\right)$$

The sensitivity of real power flow in (5.49) w.r.t. the B_i^{th} BESS output power is expressed as follows:

$$\frac{\partial P_{mn}}{\partial P_{B_i}} = \frac{1}{|V_i|}[-G_{mn}|Z_{mi}|V_{n,real}\cos\left(\theta_{mi} + \theta_i\right)$$

$$- G_{mn}|Z_{ni}|V_{m,real}\cos\left(\theta_{ni} + \theta_i\right) -$$

$$G_{mn}|Z_{mi}|V_{n,imag}\sin\left(\theta_{ni} + \theta_i\right) - G_{mn}|Z_{ni}|V_{n,imag}\sin\left(\theta_{mi} + \theta_i\right) -$$

$$B_{mn}|Z_{mi}|V_{m,real}\cos\left(\theta_{ni} + \theta_i\right) - B_{mn}|Z_{ni}|V_{n,imag}\sin\left(\theta_{mi} + \theta_i\right) +$$

$$B_{mn}|Z_{ni}|V_{n,real}\cos\left(\theta_{mi} + \theta_i\right) + B_{mn}|Z_{mi}|V_{m,imag}\sin\left(\theta_{ni} + \theta_i\right)]$$

The sensitivity of real power flow in (5.49) w.r.t. the jth DFIG disturbance power injection is expressed as follows:

$$\frac{\partial P_{mn}}{\partial P_j} = \frac{1}{|V_j|}[-G_{mn}|Z_{mj}|V_{n,real}\cos\left(\theta_{mj} + \theta_j\right)$$

$$- G_{mn}|Z_{nj}|V_{m,real}\cos\left(\theta_{nj} + \theta_j\right) -$$

$$G_{mn}|Z_{mj}|V_{n,imag}\sin\left(\theta_{nj} + \theta_j\right) - G_{mn}|Z_{nj}|V_{n,imag}\sin\left(\theta_{mj} + \theta_j\right) -$$

$$B_{mn}|Z_{mj}|V_{m,real}\cos\left(\theta_{nj} + \theta_j\right) - B_{mn}|Z_{nj}|V_{n,imag}\sin\left(\theta_{mj} + \theta_j\right) +$$

$$B_{mn}|Z_{nj}|V_{n,real}\cos\left(\theta_{mj} + \theta_j\right) + B_{mn}|Z_{mj}|V_{m,imag}\sin\left(\theta_{nj} + \theta_j\right)]$$

The power injected by ith SM bus is expressed as

$$P_{ei} = E_i^0 e^{j\delta_i^0} \left[(E_i^0 e^{j\delta_i^0} - V_i e^{-j(\delta_i - \pi/2)}) |y_i| e^{j\alpha_i} \right]^*$$
$$= (E_i^0)^2 |y_i| e^{-j\alpha_i} - E_i^0 V_i^* |y_i| e^{j(\delta_i^0 + \delta_i - \pi/2 - \alpha_i)} \tag{5.50}$$

The sensitivity of power injection (5.50) w.r.t. ith SM rotor angle is expressed as follows:

$$\frac{\partial P_{ei}}{\partial \delta_i} = E_i^0 |y_i| \Sigma_{k=1 \neq i}^g |Z_{ik}||y_k| E_k^0 sin(\theta_{ki} + \delta_{ik} + \delta_{ik}^0 + \alpha_{ik}) \tag{5.51}$$

The sensitivity of power injection (5.50) w.r.t. jth SM rotor angle is expressed as follows:

$$\frac{\partial P_{ei}}{\partial \delta_j} = -E_i^0 |y_i||Z_{ij}||y_j| E_j^0 sin(\theta_{ji} + \delta_{ij} + \delta_{ij}^0 + \alpha_{ij}) \tag{5.52}$$

The sensitivity of power injection (5.50) w.r.t. n_b^{th} BESS power output is expressed as follows:

$$\frac{\partial P_{ei}}{\partial P_{B_{n_b}}} = -\frac{E_i^0 |y_i||Z_{in_b}|}{|V_{n_b}|} sin(\theta_{in_b} + \theta_{n_b} + \alpha_i - \delta_i^0 - \delta_i) \tag{5.53}$$

The sensitivity of power injection (5.50) w.r.t. power injection disturbance by n_d^{th} DFIG , i.e., P_{n_d}, can be obtained as

$$\frac{\partial P_{ei}}{\partial P_{n_d}} = -\frac{E_i^0 |y_i||Z_{in_d}|}{|V_{n_d}|} sin(\theta_{in_d} + \theta_{n_d} + \alpha_i - \delta_i^0 - \delta_i) \tag{5.54}$$

All the sensitivities make up some entries of system matrices.

5.4.1.2 MPC design with disturbance compensation

The model (5.41) and (5.42) is discretized with sampling time T_s, while doing so, the disturbance injections from DFIG and corresponding matrices are excluded. The overall discretized system is expressed as follows:

$$\mathbf{x}_d(k+1) = \mathbf{A}_d \mathbf{x}_d(k) + \mathbf{B}_d \mathbf{u}_d(k) + \mathbf{d}_d(k) \tag{5.55}$$

$$\mathbf{y}_d(k) = \mathbf{C}_d \mathbf{x}_d(k) \tag{5.56}$$

where $\mathbf{A}_d = e^{\mathbf{A}_c T_s} \in \mathbb{R}^{n_t \times n_t}$, $\mathbf{B}_d = \int_0^{T_s} e^{\mathbf{A}_c \tau} \mathbf{B}_c d\tau \in \mathbb{R}^{n_t \times m}$, $\mathbf{d}_d = \int_0^{T_s} e^{\mathbf{A}_c \tau} \mathbf{d}_c d\tau$, $\mathbf{C}_d = \mathbf{C}_c \in \mathbb{R}^{q \times n_t}$.

The incremental form of the discretized system (5.55) and (5.56) is expressed as follows:

$$\Delta \mathbf{x}_d(k+1) = \mathbf{A}_d \Delta \mathbf{x}_d(k) + \mathbf{B}_d \Delta \mathbf{u}(k) + \Delta \mathbf{d}_d(k) \tag{5.57}$$

$$\mathbf{y}_d(k+1) - \mathbf{y}_d(k) = \mathbf{C}_d(\mathbf{x}_d(k+1) - \mathbf{x}_d(k)) \tag{5.58}$$

Considering $\mathbf{r} \in \mathbb{R}^q$, the output reference comes from the system operator. The deviation between output reference and present output, in the form of an error vector, is

$$\mathbf{e} = \mathbf{y}_d - \mathbf{r}. \qquad (5.59)$$

The extended model of the system from (5.57) to (5.59) is expressed as follows:

$$\mathbf{x}(k+1) = \mathbf{A}\mathbf{x}(k) + \mathbf{B}\Delta\mathbf{u}(k) + \mathbf{D}\Delta\mathbf{d}_d(k) \qquad (5.60)$$

$$\mathbf{y}(k) = \mathbf{C}\mathbf{x}(k) \qquad (5.61)$$

where $\mathbf{x}(k) = [\Delta\mathbf{x}_d(k)\mathbf{e}(k)]^T$, $\mathbf{y} = \mathbf{e}$, $\mathbf{A} = \begin{bmatrix} \mathbf{A}_d & \mathbf{0} \\ \mathbf{C}_d\mathbf{A}_d & \mathbf{I} \end{bmatrix} \in \mathbb{R}^{(n+q)\times(n+q)}$, $\mathbf{D} = \begin{bmatrix} \mathbf{I} \\ \mathbf{0} \end{bmatrix}$,

$\mathbf{C} = [\mathbf{0} \ \mathbf{I}] \in \mathbb{R}^{(q)\times(n+q)}$, $\mathbf{B} = \begin{bmatrix} \mathbf{B}_d \\ \mathbf{C}_d\mathbf{B}_d \end{bmatrix}^T \in \mathbb{R}^{(n+q)\times(m)}$. The matrices $\mathbf{I}$, $\mathbf{0}$ are identity, zero matrices with suitable dimensions, $\mathbf{D}\Delta\mathbf{d}_d \in W$ is the disturbance term.

Disturbance compensation technique [14]
The entries of $\mathbf{D}\Delta\hat{\mathbf{d}}_d(k)$ must be among the elements of set W. Evaluation of $\mathbf{D}\Delta\hat{\mathbf{d}}_d(k-1)$ at $k-1$ time instance is

$$\mathbf{D}\Delta\mathbf{d}_d(k-1) = \mathbf{x}(k) - \mathbf{A}\mathbf{x}(k-1) - \mathbf{B}\Delta\mathbf{u}(k-1) \qquad (5.62)$$

Assumption 1: *Estimated disturbance at k time instant is*

$$\mathbf{D}\Delta\hat{\mathbf{d}}_d(k) = \mathbf{D}\Delta\hat{\mathbf{d}}_d(k-1) + \varepsilon \qquad (5.63)$$

where $\varepsilon \in V$ and $V \subset W$.

The assumption mentioned just above accounts for the situation where the sampling rate of the whole system is much higher than disturbance variation. The variation in disturbance term is small, and the variation in ε is even smaller.

Model predictive control
The MPC for line power flow [3] is formulated as follows:

$$\min \sum_{i=1}^{N_p} \mathbf{x}^T(k+i|k)\mathbf{Q}\mathbf{x}(k+i|k)$$

$$+\Delta\mathbf{u}^T(k+i|k)\mathbf{R}_L\Delta\mathbf{u}(k+i|k) \qquad (5.64a)$$

subject to

$$\mathbf{x}(k+1) = \mathbf{A}\mathbf{x}(k) + \mathbf{B}\Delta\mathbf{u}(k) + \mathbf{D}\Delta\hat{\mathbf{d}}_d(k) \qquad (5.64b)$$

$$\mathbf{y}(k) = \mathbf{C}\mathbf{x}(k) \tag{5.64c}$$

$$\mathbf{P}_{C\min} \leq \mathbf{u}(k) \leq \mathbf{P}_{C\max} \tag{5.64d}$$

$$\Delta\mathbf{P}_{C\min} \leq \Delta\mathbf{u}(k) \leq \Delta\mathbf{P}_{C\max} \tag{5.64e}$$

$$\sum_{i=1}^{g} \Delta u_i(k) = 0 \tag{5.64f}$$

where k is the present sampling instant of time, the state at $(k + i)$th time instant is $\mathbf{x}(k + i|k)$, the input at $(k + i)$th time instant is $\mathbf{u}(k + i|k)$, the tracking error at $(k + i)$th time instant is $\mathbf{e}(k + i|k)$, $i = 1,\ldots, N_p$. MPC estimates the states, outputs, and tracking error from the information available at present instant k.

The control sequence for the SMs/BESS to track the line flow reference is obtained by optimizing the cost function in (5.64a). The matrice $\mathbf{Q}$ is a positive definite matrix, and the matrix $\mathbf{R}_L$ is also a positive definite matrix of compatible dimensions. SMs output and SMs output rate limitations are (5.64d) and (5.64e), respectively. The power balance constraint to avoid the frequency deviation is (5.64f). Every line in the power system has a set of SMs for its power control. The set of SMs is chosen based on evaluating the Generation Shift Factor (GSF). Hence, a few selected SMs are used for MPC design. The SMs should have sufficient maximum and minimum powers (5.64d) and a rate of variation on power (5.64e) [19] to achieve the convergence of the optimization in (5.64a). Hence a few selected SMs with enough capacity and high ramp rate are used for designing the MPC [12,13].

SMs with a combination of positive and negative values of GSF are chosen to satisfy (5.64f). Overall (5.64) is an optimization problem with various constraints and a finite-time horizon. The solution of (5.64) produces optimized control sequence $\mathbf{U}^* = (\Delta\mathbf{u}^*(k), \Delta\mathbf{u}^*(k + 1), \cdots, \Delta\mathbf{u}^*(k + N_p))$. Among all, only the first, i.e., $\Delta\mathbf{u}^*(k)$, drives the system. Hildreth's QP algorithm can produce optimal control sequence as in [21].

Remark 1: *The parametric variations, i.e., load changes and generation changes, are slow with low sampling time, i.e., $T_s = 0.2s$. of MPC. The optimization problem (5.64) requires low sampling time compared to open loop rise time. So Assumption 1 is valid [11].*

Remark 2: *Converting MPC (5.64) into multiparameter programming (mp-QP) and solving with explicit MPC [28] further reduces computation time.*

5.4.2 Stability analysis

The overall closed-loop system follows the reference by obeying the terminal condition of the form:

$$\mathbf{x}(k + N_p|k) = 0 \tag{5.65}$$

Definition 1: *The cost function (5.64a) [27], with control sequence Δu obeying all limits at kth time instant, should have a finite value to guarantee the feasibility of MPC (5.64).*

Theorem 1: *If the optimal control problem (5.64) with terminal constraint (5.65) is realizable at the first time instant, then with the receding time horizon control, the line flow converges asymptotically to reference without uncertainties and bounded set with uncertainties.*

Synchro-phasor measurements and control process

An algorithmic process is listed below to understand the working and process flow of the proposed congestion management method:

1. Step 1: The Independent System Operator (ISO) evaluates the GSF of generators and BESSs for different transmission lines which are to be monitored.
2. Step 2: Power flow in the lines is measured using the PMUs deployed in the network, which transmit the data to the Phasor Data Concentrator (PDC) and control center. For the overloaded lines, the influential generators are selected based on high GSFs.
3. Step 3: The MPC controller receives data from the ISO to generate strategic control input for re-dispatching of most significant generators and BESSs.
4. Step 4: The generators and BESSs receive the control input references from the controller and re-dispatch the generators and BESSs. Line power flow being sensitive to generation injections, start to alter the line power flow in the desired way.
5. Step 5: Repeat steps 2–4 until the congestion is resolved.

5.4.3   Test results and discussion

The line flow control with MPC in 68-Bus Test System is presented with the help of different case studies.

Case A: *Double circuit line flow control with one line outage*

In this case, one of the double circuit lines, between buses 60 and 61, is removed, and MPC maintains line flow in the other line at its thermal limit. Assuming thermal limit as 2.4 p.u., which is also equal to the reference of the line power. One line trip happens at $t = 200$ s, other line power reaches 2.6 p.u., which is above the magnitude of the line thermal limit. With MPC, the line power flow is maintained at 2.4 p.u., and the MPC successfully controls line flow for variation in the system because of line outage, as shown in Figure 5.9.

Case B: *Simultaneous control of two lines*

In this case, the basic structure of the 68 bus test system is altered. The line between bus 27 and bus 53 was removed to increase the load on the other lines of the system. The lines (54,53) and (60,61) are considered for the control. The lines (54,53) carry the power of 3.231 p.u., and lines (60,61) carry 3.801 p.u., under steady conditions of the system with line outage (27,53). Assuming that the thermal power limit of both

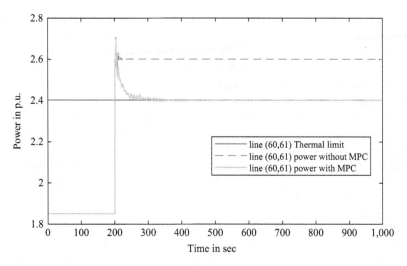

Figure 5.9 *16-Machines 68-bus test system: maintaining line power flow below the thermal limit for the tie line (60,61) after tripping of a parallel line*

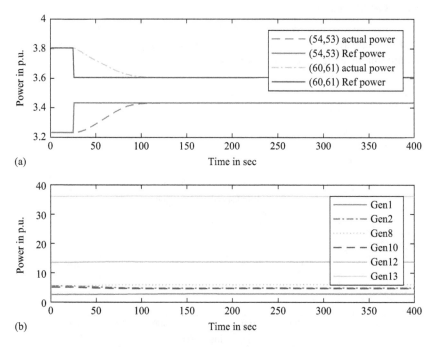

Figure 5.10 *16-Machines 68-bus test system: power flow control in tie lines (54,53) and (60,61) and power output of generators*

the lines is 3.6 p.u., MPC is activated on both lines at t = 25 s, to control the power in both lines at 3.6 p.u., and the most suitable generators for performing the control are 13, 1, 10, 2, 8, and 12 by evaluating the GSF's. The MPC is successful in controlling line flows in two different lines under overloaded condition as shown in Figure 5.10.

Case C: *Power flow stabilization with stochastic renewable power and BESS*

In this case, doubly fed induction generators (DFIGs) [24] placed at buses 30 and 57 of the test system are assumed to be stochastic power injecting sources. Rated power of the DFIG is 1 p.u. at the rated wind speed 15 m/s. The DFIGs assumed to be operating at MPPT mode and supplying optimal power. The variations in the wind speed [25] of DFIGs is very close to the realistic wind variation as shown in Figure 5.11. Any variations in the wind speed immediately reflected on output power of the DFIGs. Two BESS are placed at buses 35 and 59. For every 5 min, OPF/ED performed for SMs dispatch, MPC as a secondary control continuously suppress the load flow variations by dispatching BESSs as shown in Figure 5.12. MPC along with OPF is successful in stabilizing the power flow in (60,61) at 4.86 p.u. as shown in Figure 5.13. The SoC variations of BESS are shown in Figure 5.14.

5.5 Appendix

5.5.1 *Generator modeling*

All the SMs in the test system are modeled as classical machines, i.e., voltage behind transient reactance. As the focus is on the steady-state line flow, the effect of an automatic voltage regulator and stabilizer is negligible. Therefore, they are

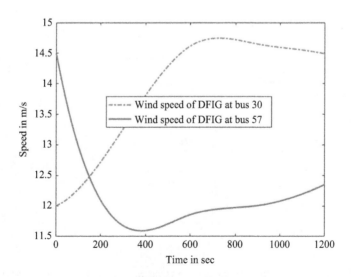

Figure 5.11 Wind speed variations for DFIGs at buses 30 and 57

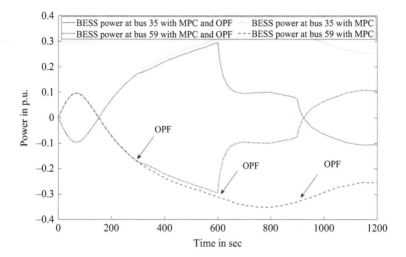

Figure 5.12   Output of BESS's placed at buses 35 and 59

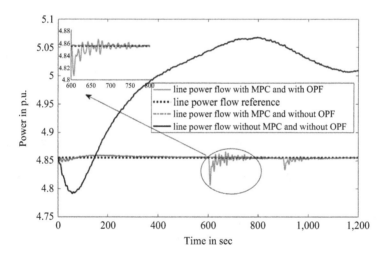

Figure 5.13   Stabilizing line power flow at initial value in tie line (60,61) in the presence of variable renewable generations

not included in the model. The voltage magnitude and angle of the classical machine are

$$E_i^0 = \sqrt{\left(\left(E_{di}^{'0}\right) + \left(X_{qi}' - X_{di}'\right)I_{qi}^0\right)^2 + \left(E_{qi}^{'0}\right)^2} \qquad (5.66)$$

$$\delta_i^0 = \tan^{-1}\left(\frac{E_{qi}^{'0}}{E_{di}^{'0} + \left(X_{qi}' - X_{di}'\right)I_{qi}^0}\right) \qquad (5.67)$$

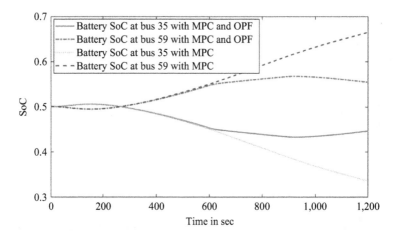

Figure 5.14   SoC of BESS's placed at buses 35 and 59

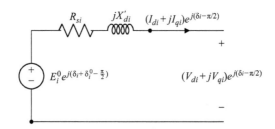

Figure 5.15   Voltage source equivalent model of the ith *generator*

The SM as voltage source [23] shown in Figure 5.15.
The rotor dynamics are as follows:

$$\frac{d\delta_i}{dt} = \omega_B(\omega_i - \omega_s) = \omega_B.\Delta\omega_i \tag{5.68}$$

$$2H_i\frac{d\omega_i}{dt} = P_{M_i} - P_{ei} - D_i(\omega_i - \omega_s) \tag{5.69}$$

$$P_{ei} = E_i^0 e^{j\left(\delta_i + \delta_i^0 - \frac{\pi}{2}\right)}\left(I_{di} - jI_{qi}\right)e^{-j\left(\delta_i - \frac{\pi}{2}\right)} \tag{5.70}$$

where $i = 1, 2, \ldots\ldots, n_g$.

5.5.2   Governor-turbine model

The governor dynamics are [23] as follows:

$$T_{SV_i}\frac{dP_{SV_i}}{dt} = -P_{SV_i} + P_{C_i} - \frac{1}{R_{D_i}}(\omega_i - \omega_s) \tag{5.71}$$

The turbine dynamics are as follows:

$$T_{CH_i} \frac{dP_{CH_i}}{dt} = -P_{CH_i} + P_{SV_i} \tag{5.72}$$

$$T_{RH_i} \frac{dP_{M_i}}{dt} = -P_{M_i} + \left(1 - \frac{K_{HP_i} T_{RH_i}}{T_{CH_i}}\right) P_{CH_i} + \frac{K_{HP_i} T_{RH_i}}{T_{CH_i}} P_{SV_i} \tag{5.73}$$

The block diagram representation of the turbine-governor system is shown in Figure 5.16.

5.5.3 BESS

BESS dynamics [22] are as follows:

$$\tau_i \frac{dP_{B_i}}{dt} = -P_{B_i} + k_i P_{B_i}^* \tag{5.74}$$

where W_i, τ_i, P_{B_i}, k_i, SOC_i^0, SOC_i are capacity of the inverter, inverter time constant, output power, gain of the controller, state of charge at initially, and state of charge ith BESS respectively. $\eta = \eta_c$ and $\eta = 1/\eta_d$ are charging and discharging efficiencies, respectively. The complete block diagram is represented in Figure 5.17.

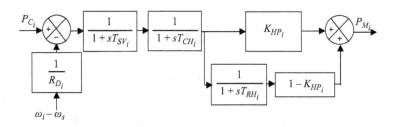

Figure 5.16   Transfer function model of the prime-mover systems

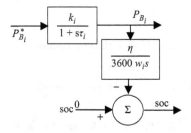

Figure 5.17   Battery energy storage system model

5.5.4   Generation shift factor

The change in the power flow from bus m to bus n is ΔP_{mn}, ΔP_k is the output change at k^{th} SM. Therefore, the GSF is defined as follows:

$$GSF_{mn,k} = \frac{\Delta P_{mn}}{\Delta P_k} \tag{5.75}$$

5.5.5   Stability proof of MPC

5.5.5.1   In the absence of uncertainty

The formulated cost function (5.64a) would have minimum value with control sequence $\Delta \mathbf{u}^*$ obtained from solving (5.64). Considering the Lyapunov function candidate as the optimal value

$$
\begin{aligned}
V(k) &= \sum_{m=0}^{N_p} \mathbf{x}^*(k+m|k)^T \mathbf{Q}\mathbf{x}^*(k+m|k) + \\
&\quad \sum_{m=0}^{N_p-1} \Delta\mathbf{u}^*(k+m|k)^T \mathbf{R}_L \Delta\mathbf{u}^*(k+m|k)
\end{aligned}
\tag{5.76}
$$

where $\mathbf{x}^*(k|k) = \mathbf{x}(k)$ and

$$
\begin{aligned}
\mathbf{x}(k+m|k) &= \mathbf{A}^m \mathbf{x}(k) + \mathbf{A}^m \mathbf{B}\Delta\mathbf{u}(k) \\
&\quad + \cdots + \mathbf{A}\mathbf{B}\Delta\mathbf{u}(k+m-1)
\end{aligned}
\tag{5.77}
$$

At $(k+1)$th time instant, (5.76) rewritten as

$$
\begin{aligned}
V(k+1) &= \sum_{m=1}^{N_p+1} \mathbf{x}^*(k+m+1|k+1)^T \mathbf{Q}\mathbf{x}^*(k+m+1|k+1) \\
&\quad + \sum_{m=1}^{N_p} \Delta\mathbf{u}^*(k+m+1|k+1)^T \mathbf{R}_L \Delta\mathbf{u}^*(k+m+1|k+1)
\end{aligned}
\tag{5.78}
$$

At $(k+1)$th time instant, (5.64) is feasible only if the generated control sequence, i.e., $\Delta\mathbf{u}(k+N_p|k+1)$ at $(k+1)$th time instant satisfies the following equation:

$$\mathbf{x}(k+N_p+1|k+1) = 0 \tag{5.79}$$

A positive definite function at $(k+1)$th time instant defined as

$$
\begin{aligned}
\overline{V}(k+1) &= \sum_{m=1}^{N_p} \mathbf{x}^*(k+m|k+1)^T \mathbf{Q}\mathbf{x}^*(k+m|k+1) \\
&\quad + \sum_{m=1}^{N_p-1} \Delta\mathbf{u}^*(k+m|k)^T \mathbf{R}_L \Delta\mathbf{u}^*(k+m|k) \\
&\quad + \Delta\mathbf{u}(k+N_p|k+1)^T \mathbf{R}_L \Delta\mathbf{u}(k+N_p|k+1)
\end{aligned}
\tag{5.80}
$$

Equation (5.80) is obtained from (5.78) by altering it with a feasible control sequence. $V(k+1)$ is upper bounded by $\overline{V}(k+1)$, and the following inequality holds true

$$V(k+1) - V(k) \leq \overline{V}(k+1) - V(k) \tag{5.81}$$

For time instances, $k+1, k+2, \cdots, k+N_p-1$, the function $\overline{V}(k+1)$ and $V(k)$ share the same sequence of control signal, and the difference would be as follows:

$$\begin{aligned}\overline{V}(k+1) - V(k) = \mathbf{x}^{*T}(k+N_p|k)\mathbf{Q}\mathbf{x}^{*}(k+N_p|k) - \\ \mathbf{x}^{*}(k)^T\mathbf{Q}\mathbf{x}^{*}(k) - \Delta\mathbf{u}^{*}(k)^T\mathbf{R}_L\Delta\mathbf{u}^{*}(k).\end{aligned} \tag{5.82}$$

Since the terminal constraint is zero. The upper bound on the $V(k+1) - V(k)$ is as follows:

$$\overline{V}(k+1) - V(k) = -\mathbf{x}^{*T}(k)\mathbf{Q}\mathbf{x}^{*}(k) - \Delta\mathbf{u}^{*T}(k)\mathbf{R}_L\Delta\mathbf{u}^{*}(k). \tag{5.83}$$

Let $\lambda_{\min}$ be the minimum eigenvalue of the matrix Q, then using the relations (5.39) and (5.41), we obtain:

$$V(k+1) - V(k) \leq -\lambda_{\min}(\mathbf{Q})\|\mathbf{x}^{*}(k)\|^2 - \mathbf{R}_L\|\Delta\mathbf{u}^{*}(k)\|^2. \tag{5.84}$$

Therefore, the considered Lyapunov function is asymptotically stable. Hence, the augmented state and error asymptotically converge to zero.

References

[1]   U. Samanmit, S. Chusanapiputt, and V. Pungprasert, "Increasing of dynamic thermal rating of transmission line," in: *2006 International Conference on Power System Technology*, 2006, pp. 1–4, doi: 10.1109/ICPST.2006.321536.

[2]   F. U. Haq, P. Bhui, and K. Chakravarthi, "Real time congestion management using plug in electric vehicles (PEV's): a game theoretic approach," *IEEE Access*, vol. 10, pp. 42029–42043, 2022, doi: 10.1109/ACCESS.2022.3167847.

[3]   K. Chakravarthi, P. Bhui, N. K. Sharma, and B. C. Pal, "Real time congestion management using generation re-dispatch: modeling and controller design," *IEEE Transactions on Power Systems*, vol. 10, pp. 42029–42043, 2022, doi: 10.1109/TPWRS.2022.3186434.

[4]   C. Kotakonda and P. Bhui, "Modelling of power flow dynamics for congestion control," in: *2022 IEEE International Conference on Signal Processing, Informatics, Communication and Energy Systems (SPICES)*, 2022, pp. 320–325, doi: 10.1109/SPICES52834.2022.9774150.

[5]   C. Chen, S. V. Dhople, A. D. Dom ınguez-Garc ıa, and P. W. Sauer, "Generalized injection shift factors," *IEEE Transactions on Smart Grid*, vol. 8, no. 5, pp. 2071–2080, 2016.

[6]   S. Pal, S. Sen, and S. Sengupta, "Power network reconfiguration for congestion management and loss minimization using Genetic Algorithm," *IET Conference Publications*, 2015, pp. 291–296, doi:10.1049/cp.2015.1646.

[7]   M. Khanabadi and H. Ghasemi, "Transmission congestion management through optimal transmission switching," in *2011 IEEE Power and Energy Society General Meeting*, IEEE, 2011, pp. 1–5.

[8]   S. Dutta and S. P. Singh, "Optimal rescheduling of generators for congestion management based on particle swarm optimization," *IEEE Transactions on Power Systems*, vol. 23, no. 4, pp. 1560–1569, 2008.

[9]   W. Liu, Q. Wu, F. Wen, and J. Østergaard, "Day-ahead congestion management in distribution systems through household demand response and distribution congestion prices," *IEEE Transactions on Smart Grid*, vol. 5, no. 6, pp. 2739–2747, 2014.

[10] J. Zhang and A. Yokoyama, "Optimal power flow control for congestion management by interline power flow controller (ipfc)," in: *2006 International Conference on Power System Technology*, IEEE, 2006, pp. 1–6.

[11] R. Patel, L. Meegahapola, L. Wang, X. Yu, and B. McGrath, "Automatic generation control of multi-area power system with network constraints and communication delays," *Journal of Modern Power Systems and Clean Energy*, vol. 8, no. 3, pp. 454–463, 2020, doi: 10.35833/MPCE.2018.000513.

[12] J. S. A. Carneiro and L. Ferrarini, "Preventing thermal overloads in transmission circuits via model predictive control," *IEEE Transactions on Control Systems Technology*, vol. 18, no. 6, pp. 1406–1412, 2010, doi: 10.1109/TCST.2009.2037921.

[13] J. A. Martin and I. A. Hiskens, "Corrective model-predictive control in large electric power systems," *IEEE Transactions on Power Systems*, vol. 32, no. 2, pp. 1651–1662, 2017, doi: 10.1109/TPWRS.2016.2598548.

[14] Z. Li and J. Sun, "Disturbance compensating model predictive control with application to ship heading control," *IEEE Transactions on Control Systems Technology*, vol. 20, no. 1, pp. 257–265, 2012, doi: 10.1109/TCST.2011.2106212.

[15] A. Asrari, M. Ansari, J. Khazaei, and P. Fajri, "A market framework for decentralized congestion management in smart distribution grids considering collaboration among electric vehicle aggregators," *IEEE Transactions on Smart Grid*, vol. 11, no. 2, pp. 1147–1158, 2019.

[16] L. Zhang and Y. Li, "A game-theoretic approach to optimal scheduling of parking-lot electric vehicle charging," *IEEE Transactions on Vehicular Technology*, vol. 65, no. 6, pp. 4068–4078, 2015.

[17] P. Bhui and N. Senroy, "Online identification of tripped line for transient stability assessment," *IEEE Transactions on Power Systems*, vol. 31, no. 3, pp. 2214–2224, 2016, doi: 10.1109/TPWRS.2015.2440562.

[18] P. Bhui and N. Senroy, "Real-time prediction and control of transient stability using transient energy function," *IEEE Transactions on Power Systems*, vol. 32, no. 2, pp. 923–934, 2017, doi: 10.1109/TPWRS.2016.2564444.

[19] Z. Sun, Y. Liu, J. Wang, *et al.*, "Applications of game theory in vehicular networks: a survey," *IEEE Communications Surveys Tutorials*, vol. 23, no. 4, pp. 2660–2710, Fourth quarter 2021, doi: 10.1109/COMST.2021.3108466.

[20] H. S. V. S. K. Nunna, A. Sesetti, A. K. Rathore, and S. Doolla, "Multiagent-based energy trading platform for energy storage systems in distribution systems with interconnected microgrids," *IEEE Transactions on Industry Applications*, vol. 56, no. 3, pp. 3207–3217, 2020, doi: 10.1109/TIA.2020.2979782.

[21] C. Hildret, "A quadratic programming procedure," *Naval Research Logistics Quarterly*, vol. 4, pp. 79–85, 1957.

[22] M. Toge, Y. Kurita, and S. Iwamoto, "Supplementary load frequency control with storage battery operation considering SOC under large-scale wind power penetration," in: *2013 IEEE Power Energy Society General Meeting*, 2013, pp. 1–5, doi: 10.1109/PESMG.2013.6672323.

[23] P. Kundur, *Power System Stability and Control*, New York, NY: McGraw-Hill, 1994.

[24] M. S. Patel and N. N. Shah, "Small signal stability of doubly fed induction generator connected to the grid," in: *2019 3rd International Conference on Trends in Electronics and Informatics (ICOEI)*, 2019, pp. 473–477, doi: 10.1109/ICOEI.2019.8862549.

[25] M. Yu, W. Zhou, B. Wang, and J. Jin, "The short-term forecasting of wind speed based on EMD and ARMA," in: *2017 12th IEEE Conference on Industrial Electronics and Applications (ICIEA)*, 2017, pp. 495–498, doi: 10.1109/ICIEA.2017.8282895.

[26] CERC, Report on the Grid Disturbances on 30th July and 31st July 2012. Available at: http://www.cercind.gov.in, 2012.

[27] J. B. Rawlings, D. Q. Mayne, and M. Diehl, *Model Predictive Control: Theory, Computation, and Design*, San Francisco, CA: Nob Hill Publishing, 2017.

[28] A. Bemporad, M. Morari, V. Dua, and E. N. Pistikopoulos, "The explicit linear quadratic regulator for constrained systems," *Automatica*, vol. 38, no. 1, pp. 3–20, 2002. Available at: https://www.sciencedirect.com/science/article/pii/S0005109801001741.

Chapter 6

Out-of-step predictive wide-area protection

Majid Sanaye-Pasand[1] and Sayyed Mohammad Hashemi[2]

6.1 Introduction

Protection schemes are designed and deployed to either protect the power system components from being damaged against various types of faults, e.g. short-circuits or preserve the integrity and stability of power system against large disturbances. These critical functions have traditionally been implemented in practice based on the local measurements. This way, the action of protection system is mainly corrective, rather than preventive.

Wide-area measurements can enhance the performance of protection system by providing a broader view from the power system condition. The protective relay can now access many critical parameters measured from different parts of the grid. This is particularly useful to the system protection schemes whose action may affect a large area of the power system. Wide-area measurements help the relay to make a better decision on which parts of the systems and when those sections should be tripped. They also facilitate taking several pictures from the past and present operating conditions of power system, which can be used in the prediction of the future states of the system. Therefore, wide-area protection scheme can take predictive actions to prevent possible faults or power outages to happen [1].

An appropriate candidate for application of wide-area measurements in power system protection is the out-of-step (OOS) relaying of transmission lines. When an area of power system experiences unstable angular swings with respect to the rest of the grid, there would be a group of transmission lines that can be tripped to rescue the whole system from the risk of angle instability [2,3]. The faster those lines are tripped, the more the stability margin is. This action can be carried out either correctively, i.e. once the OOS happened, or predictively, i.e. before the evolution of OOS.

Conventional OOS protection is based on the impedance trajectory seen by distance relays. The changes in the source impedance, system operating condition, and network topology may affect the performance of conventional OOS protection [4]. An effective alternative solution is deploying WAMS system to implement a

[1]School of Electrical and Computer Engineering, College of Engineering, University of Tehran, Iran
[2]Office of Grid Protection, Iran Grid Management Company (IGMC), Iran

wide-area OOS protection. In this regard, OOS can be detected using the equal area criterion (EAC) which is a fundamental transient stability assessment tool. EAC is basically presented for a single-machine system, but it can be extended to the multi-machine systems, as well. This can be done by the single-machine infinite-bus modeling and network reduction, based on the online wide-area measurement of electrical signals at the generator buses. Wide-area OOS protection can also be carried out by using the state-plane trajectories [5], decision tree-based techniques [6], and apparent impedance trajectory [7].

Taking some steps back to predict the OOS before its occurrence would help to save a considerable time which can be used to take more effective remedial actions. From the technical and economical viewpoints, countermeasures such as preplanned line/generator tripping and intentional islanding are much better than unpredicted and uncontrolled outages of system components due to the OOS outcomes. The prediction can be achieved by available mathematical tools and/or heuristic methods in the model-based or curve-fitting-based forms [8]. Model-based prediction relies on the modeling of system dynamics by mathematical functions and control blocks. It needs to know the dynamic characteristics of system components and changes from one system to another. On the other hand, curve-fitting-based prediction is a pure mathematical tool to reveal the trend of time series, e.g. rotor angle trajectory, and does not need a prior knowledge about the dynamic model of system components.

In this chapter, a new approach for the prediction of OOS using wide-area measurements is presented. The method predicts the future samples of power and angle curves by a time-series modeling technique called the auto-regressive moving-average (ARMA) method. The conventional version of ARMA is not suitable for online applications, due to the heavy computation burden. To cope with this shortcoming, an extended Kalman filtering (EKF) technique is used for estimating the parameters of ARMA model. EKF can also improve the efficiency of ARMA modeling by attenuating the effect of noisy measurements. It also enhances the robustness of proposed method against missing PMU data. The enhanced ARMA/EKF method is developed for predicting the power and angle curves. The OOS conditions in single generators and multi-machine power systems are detected by EAC and extended EAC (EEAC), respectively. Depending on the PMUs' reporting and the number of predicted samples, the proposed method can predict OOS in hundreds of milliseconds before its actually occurs. This time is long enough to activate appropriate countermeasures against transient instability and its spreading over the grid.

6.2 Analytic prediction vs. artificial intelligent-based prediction

Wide-area measurements in a power system often provide large-size high-dimension data sets, which are updated continuously. They can be treated mathematically as time series. A time series is a series of observations or data, usually taken at equal time intervals, which represents an underlying stochastic process of the system under observation. Time series forecasting has been a topic of interest in many disciplines since the decades ago. It involves predicting the next observation of the time series based on known previous observations at each step.

A traditional approach to time series forecasting is ARMA modeling. This is actually a statistical analysis-based approach, which reveals the correlation between the past and current observations. As mentioned earlier, ARMA modeling can be considered as a curve-fitting-based prediction method and does not need a prior knowledge about system dynamics. Nevertheless, there are more analytic approaches to predict OOS and transient instability of power systems. For example, in [9], the speed–acceleration curve of the two-machine equivalent model of the system is extracted and its slope is used as a criterion for inter-area OOS prediction. In [10], a reduced order linear model of the system is estimated utilizing wide-area measurements and then, the system stability is predicted in a sliding prediction window using the relative rotor angle. The method presented in [11] employs the rate of change of positive-sequence voltage angle at the instant of critical clearing time as an indicator for predicting if the particular generator advances to an OOS condition.

Beyond the analytic methods, artificial intelligence (AI)-based methods have also been developed to forecast time series and predict power system transient instability. AI includes a wide range of training-based methods, from machine learning to deep learning, decision tree, and artificial neural networks. Deep learning neural networks can automatically learn arbitrary complex mappings from inputs to outputs and offer a lot of promise for time series forecasting, particularly on problems with complex-nonlinear dependencies, multivalent inputs, and multi-step forecasting. In [12,13], two deep learning-based solutions are proposed for wide-area OOS prediction. Reference [14] introduces a decision tree based for the prediction of transient instability in power systems. The analytic and AI-based methods are combined in [15] in the form of trajectory fitting and extreme learning machine to effectively predict the transient stability. AI-based methods are fast enough for online applications. However, they suffer from requiring to large training sets and extensive simulation studies.

6.3 Prediction of power and angle curves by time series modeling

In this section, concepts of ARMA and EKF methods are reviewed in brief. Then, EKF is adopted to estimate the coefficients of ARMA model. Afterwards, prediction of power and angle curves by the proposed ARMA/EKF approach, along with some practical considerations, is discussed.

A. Time series modeling using ARMA

Time series, as a sequence of data points versus time, can be modeled by several methods. Modeling of time series helps unveil and predict its general trend and future variations. The data points may be correlated and have some degrees of dependency to each other. The complex relationship between the current and previous data points can be unveiled by statistical analytic tools. Most of the practical time series follow the ARMA model, which describes the current data point of a time series as a linear combination of previous data points and errors. As the number of previous data points (p) and errors (q) increase, the complexity of ARMA model and its order increase as well. The ARMA model can be expressed mathematically as [16]:

$$\hat{y}(k) = a_1 y(k-1) + \cdots + a_p y(k-p) + c_0 e(k) + \cdots c_q e(k-q) \tag{6.1}$$

Equation (6.1) describes the time series y as an ARMA(p,q) model, which predicts the unobserved data point $\hat{y}(k)$ based on the observed data set $\{y(k-1), y(k-2), \ldots, y(k-p)\}$ and the error data set $\{e(k), e(k-1), \ldots, e(k-q)\}$. The coefficients $\{a_1, a_2, \ldots, a_p\}$ and $\{c_0, c_1, \ldots, c_q\}$ in (6.1) should be found such that the mean square error is minimized [16]. There are some solutions for the calculation of ARMA model coefficients, such as the Yule–Walker, Burg and Hannan–Rissanen algorithms. However, they have a quite heavy computation burden and are rarely used for online calculations [17].

B. Estimation by extended Kalman filtering algorithm

Kalman filtering algorithm is mainly used to estimate unknown variables in a series of measurements observed over time based on a joint probability distribution of the variable measurements. The basic recursive discrete-time extended Kalman filter (EKF) is described as [18]:

$$\begin{cases} \mathbf{X}(k+1) = \mathbf{A}(k)\mathbf{X}(k) + \mathbf{W}(k) \\ \mathbf{Y}(k) = \mathbf{C}(k)\mathbf{X}(k) + \mathbf{V}(k) \end{cases} \tag{6.2}$$

where $\mathbf{X}$ is the $n \times 1$ system states matrix, $\mathbf{A}$ is the $n \times n$ time-varying state transition matrix, $\mathbf{Y}$ is the $m \times 1$ measurement vector, $\mathbf{C}$ is the $m \times n$ time-varying output matrix, $\mathbf{W}$ is the $n \times 1$ system error matrix, and $\mathbf{V}$ is the $m \times 1$ measurement error matrix. $\mathbf{W}(k)$ and $\mathbf{V}(k)$ are white noises that have zero means and no time correlation. The incorporated covariance matrices are defined as:

$$\begin{aligned} E[\mathbf{W}(k)\mathbf{W}^T(k)] &= \mathbf{Q}(k) \\ E[\mathbf{V}(k)\mathbf{V}^T(k)] &= \mathbf{R}(k) \end{aligned} \tag{6.3}$$

where E is the expected value. The EKF algorithm includes three calculation steps, as the following set of equations.

Kalman gain:

$$\mathbf{K}(k) = [\mathbf{A}(k)\mathbf{P}(k)\mathbf{C}^T(k)]\,[\mathbf{C}(k)\mathbf{P}(k)\mathbf{C}^T(k) + \mathbf{R}(k)]^{-1} \tag{6.4}$$

New state estimate:

$$\mathbf{X}(k+1) = \mathbf{A}(k)\mathbf{X}(k) + \mathbf{K}(k)[\mathbf{Y}(k) - \mathbf{C}(k)\mathbf{X}(k)] \tag{6.5}$$

Error covariance update:

$$\mathbf{P}(k+1) = [\mathbf{A}(k) - \mathbf{K}(k)\mathbf{C}(k)]^T \mathbf{P}(k)[\mathbf{A}(k) - \mathbf{K}(k)\mathbf{C}(k)] + \mathbf{K}(k)\mathbf{R}(k)\mathbf{K}^T(k) \tag{6.6}$$

To improve the convergence characteristic of Kalman filter, the initial state value is calculated as:

$$X(0) = [H^T R^{-1} H]^{-1} H^T R^{-1} Y(0)$$
$$P(0) = [H^T R^{-1} H]^{-1}$$

(6.7)

where $Y(0) = [y(1), y(2), \ldots, y(m)]^T$ is the vector of m measured samples and $H = [c(1), c(2), \ldots, c(m)]^T$.

C. Estimation of ARMA model parameters by EKF

The proposed EKF parameter estimator reduces the computation burden of ARMA modeling and alleviates the adverse effects of noisy measurements. It also enhances the power/angle prediction model when some PMU data are missed or corrupted. To estimate the coefficients of ARMA model by EKF, they are replaced in matrix X of (6.2) and (6.5) as [19]:

$$X = \begin{bmatrix} a_1 & a_2 & \cdots & a_p & c_0 & c_1 & \cdots & c_q \end{bmatrix}^T$$

(6.8)

In each data window, containing $p+q$ samples, the ARMA coefficients or the states of EKF are fixed and time independent. It means that the states do not change in time, the state transition matrix A is a constant $(p + q) \times (p + q)$ identity matrix. If there is no random input and measurements are noise-free, W and V become zero. Otherwise, they can be assumed as a probability distribution of random numbers to consider the noise effects. Matrix C in (6.2) and (6.5) is built as:

$$C = [y(k - 1) \quad \cdots \quad y(k - p) \quad e(k) \quad \cdots \quad e(k - q)]$$

(6.9)

In each data window, for a given step k, once the matrices X and C are found, the predicted value of time series is calculated as $Y(k) = C(k)X(k)$.

D. Prediction of power and angle curves

One of the well-proven tools for transient stability assessment is the equal area criterion. It works based on the areas created by the power and angle curves in the pre- and post-disturbance conditions. Therefore, the prediction of power and angle curves would help to predict the state of stability before it happens in the reality. Prediction of future curves needs to measure the current and past time samples of the curves, which can be achieved by wide-area measurement systems. At each generator bus, the power and angle are measured by a dedicated PMU and communicated to the control center. Theoretically, the prediction can be performed for any number of future samples. However, the larger the number of predicted samples, the lower the accuracy of prediction. Therefore, the number of predicted samples and the accuracy of prediction should be compromised.

As a numerical example, Figure 6.1 shows a typical angle curve during stable power swings, starting at $t = 1$ s. The actual and ideal angle curves are indicated in blue and gray, respectively. Mathematically, the ideal predicted curve y_{id} can be

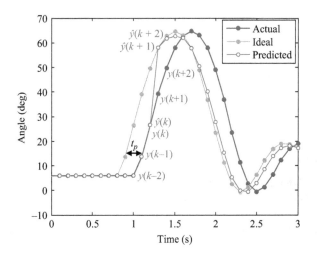

Figure 6.1   A typical angle curve during stable power swings

interpreted as an actual measured curve which is shifted in time, or:

$$y_{id}(k - n_p) = y(k) \tag{6.10}$$

where n_p is the number of samples during the shift time t_p. The ideal predicted curve is used as a benchmark to evaluate the accuracy of prediction. Indeed, the objective of prediction is to follow the ideal curve.

The prediction of the angle curve by the proposed ARMA/EKF would result in the red curve shown in Figure 6.1, where the order of ARMA(p,q) model is (2,2). For the first data window, the prediction starts at the second sample after the disturbance inception instant and proceeds two samples ahead. Each data window includes both actual and predicted time samples, the numbers of which are determined by the order of ARMA and desired prediction horizon, respectively.

The data window used in Figure 6.1 includes two actual and three predicted samples. First, sample $\hat{y}(k)$ is predicted using a linear combination of $y(k-1)$, $y(k-2)$ and their incorporated errors. Then, $\hat{y}(k + 1)$ is predicted using $\hat{y}(k)$, $y(k - 1)$ and their errors. Finally, the same process is applied to predict $\hat{y}(k + 2)$ by using $\hat{y}(k + 1)$ and $\hat{y}(k)$. The whole process of prediction is updated sample-by-sample once the new samples of actual curve enter the data window.

Compared to the ideal curve, the predicted curve shown in Figure 6.1 is accompanied with small errors, which basically depends on the number of prediction steps and sampling frequency. Effects of these parameter on the quality of prediction can be elaborated by using the R-squared (R^2) index as goodness-of-fit (GOF) criteria. This index is defined as [20]:

$$R^2 = 1 - \frac{\sum_{k=1}^{K} [y_{id}(k) - \hat{y}(k)]^2}{\sum_{k=1}^{K} [y_{id}(k) - \bar{y}_{id}]^2} \tag{6.11}$$

where K is the total number of samples within the evaluation interval. An example of prediction evaluation is presented here for the following angle curve:

$$\delta(t) = 30e^{-0.1t} \sin(2\pi f_s t) \tag{6.12}$$

In (6.12), f_s is the swing frequency and assumed to be 1 Hz. Table 6.1 shows the results of prediction evaluation, where F_r and Δt denote the sampling frequency and time difference between samples, respectively. According to the IEEE Std. C37.118, PMUs in 50 Hz systems shall support the reporting rates (F_r) of 10, 25, and 50 Hz [21]. These reporting rates (sampling frequencies) are considered in Table 6.1 to assess the accuracy of prediction for some samples ahead. For a given F_r, as much as the number of predicted samples (n_p) increases, a longer time (t_p) can be saved. However, this increase is usually achieved at the expense of losing the accuracy.

Generally speaking, by using a limited number of the measured samples, it is not possible to predict so many future samples. For high values of F_r, the number of predicted samples n_p should be increased to save more time $t_p = n_p/F_r$. Thus, the window length and the order of ARMA model should be adapted with the required t_p. Moreover, some important transient variations, especially in the power curve, will be lost, if the reporting rate is low.

Fault clearing time in transmission systems is usually fast and less than 100 ms. For low reporting rates, it would not be possible to calculate the accelerating area used in EAC during this short time and all disturbances with the accelerating areas about 100 ms may be detected as stable. This is not a concerning issue in practice, since the critical clearing time (CCT) in large and interconnected power systems is usually more than 400 ms and the probability of OOS for short clearing times is very low. As a result, the reporting rate and number of predicted samples should be set considering the CCT of the system.

Table 6.1 Evaluation of prediction accuracy

ARMA parameters		Measurement parameters		Prediction parameters		GOF index
p	q	F_r (Hz)	Δt (s)	n_p	t_p (ms)	R^2
5	10	50	0.02	10	200	0.9837
10	5	50	0.02	10	200	0.9696
20	30	50	0.02	15	300	0.9801
10	10	50	0.02	15	300	0.9774
30	30	50	0.02	20	400	0.9736
3	7	25	0.04	5	200	0.8986
10	15	25	0.04	5	200	0.9344
10	15	25	0.04	8	320	0.8937
15	20	25	0.04	8	320	0.9002
2	2	10	0.1	2	200	0.5730
4	4	10	0.1	2	200	0.5766
2	2	10	0.1	3	300	0.5080
4	4	10	0.1	3	300	0.6146
3	3	10	0.1	4	400	0.5154

6.4   Proposed OOS prediction algorithm

6.4.1   General description

As mentioned earlier, the proposed predictive approach for wide-area OOS protection is based on EAC. It can be easily implemented using the electrical power
and rotor angle measured by the PMU installed at the generator terminal. For a
single generator, OOS can be detected using the basic form of EAC, while for the
inter-area OOS detection, a single-machine infinite bus (SMIB) model of the system should be built based on the generators clustering and equivalent power and
angle at the center of angle. Then, the stability of generators can be assessed using
the extended EAC. This process will be explained in the next part.

One of the basic requirements of the proposed method is that all generator-
connected buses must be directly monitored by dedicated PMUs, which can be
easily met in modern power systems. The power and angle can be either directly
communicated to the control center or calculated in the control center from the
reported voltage and current phasors. Thanks to the advancements in the communication technologies, the proposed method is already straightforward to be
implemented in practice.

6.4.2   Details

As shown in Figure 6.2, the proposed algorithm for wide-area OOS protection is
triggered by the occurrence of abnormal conditions. During normal condition, all
generators are synchronous with each other and for each one, the electrical and
mechanical powers are equal. Following a large disturbance, e.g. a three-phase fault
near the generator, the electrical power (P_e) of generator becomes less than the
mechanical power (P_m). The proposed algorithm utilizes the difference between the
electrical and mechanical powers (ΔP_e) as an index for abnormal condition detection. Once ΔP_e becomes less than -0.1 p.u. at any generator bus, an abnormal
condition is detected and the proposed algorithm is started.

After that, to extract the SMIB model of the system, the coherent generators
are identified and clustered in the critical and remaining groups. Identification of
coherent generators can be performed either by using offline (model-based) or
online (measurement-based) techniques. The offline method uses the structural
information of power system, such as the generators locations and appropriate
splitting boundaries, to determine the controlled islands and coherent generators.
On the other hand, the online method identifies the coherent generators by
revealing the statistical similarities between their rotor angle or rotor speed variations. In the power systems where the majority of generation is concentrated in a
few specific area, the clusters are predefined the controlled island can be easily
formed via the tie lines [22–25]. Reference [26] presents a comprehensive critical
cluster identification method based on the generators critical clearing times.

The proposed algorithm employs an online critical clustering method based on
the change of angular speeds of generators which gives satisfactory results: a given
generator *i* is assigned to the critical cluster if the absolute value of its angular

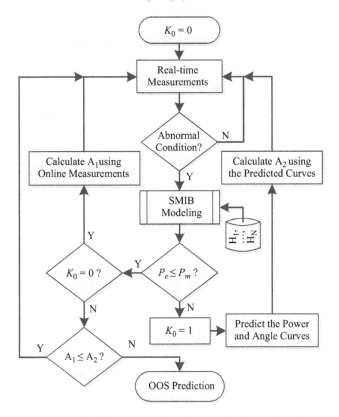

Figure 6.2   Proposed algorithm for OOS prediction

speed deviation is more than the average speed deviation of all N generators, or:

$$|\Delta\omega_i(t)| > SF \times \frac{1}{N} \sum_{i=1}^{N} |\Delta\omega_i(t)| \tag{6.13}$$

where SF is the safety factor (1.1~2) and $\Delta\omega$ can be either measured online or calculated numerically by differentiating the measured rotor angle.

Once the critical and remaining clusters are identified, the SMIB model is ready to build. Denoting the critical and remaining clusters by S and A, respectively, the equivalent inertia M is calculated for each cluster as:

$$M_S = \sum_{k\in\{S\}} M_k, \qquad M_A = \sum_{k\in\{A\}} M_k \tag{6.14}$$

The equivalent rotor angle for each cluster is obtained as:

$$\delta_S = \frac{1}{M_S} \sum_{k\in\{S\}} M_k\delta_k, \qquad \delta_A = \frac{1}{M_A} \sum_{k\in\{A\}} M_k\delta_k \tag{6.15}$$

For each cluster, the swing equation is expressed as:

$$M_S \frac{d^2 \delta_S}{dt^2} = \sum_{k \in \{S\}} (P_{mk} - P_{ek}), \qquad M_A \frac{d^2 \delta_A}{dt^2} = \sum_{k \in \{A\}} (P_{mk} - P_{ek}) \qquad (6.16)$$

The two-machine equivalent model of the system is established using (6.14) to (6.16). To reduce the two-machine model to an SMIB model, the following set of equations is used:

$$\begin{cases} \delta = \delta_A - \delta_S \\ M = \dfrac{M_A M_S}{(M_A + M_S)} \\ P_m = \dfrac{(M_S P_{mA} - M_A P_{mS})}{M_A + M_S} \\ P_e = \dfrac{(M_S P_{eA} - M_A P_{eS})}{M_A + M_S} \end{cases} \qquad (6.17)$$

where δ, M, P_m, and P_e are the equivalent angle, inertia, mechanical power, and electrical power of the system, respectively.

The power and angle curves are directly measured until the fault is cleared. Once the fault clearance is detected, the algorithm waits for the minimum required samples to enter into the data window. The waiting time depends on the order of ARMA model. Afterwards, the power and angle curves are predicted and finally, the EAC is used to check whether the predicted curves would result in OOS or not. To do so, the accelerating (A_1) and decelerating (A_2) areas are calculated as

$$\begin{aligned} A_1 &= \sum [P_m(k) - P_e(k)] \times \Delta\delta \\ A_2 &= \sum [P_e(k) - P_m(k)] \times \Delta\delta \end{aligned} \qquad (6.18)$$

where k is the sample number. The OOS is detected by EAC when the decelerating area A_2 is smaller than the accelerating area A_1. The control parameter K_0 changes from 0 to 1 once A_2 is completely calculated. As noted in Figure 6.2, A_1 is calculated using online measurements. The prediction of power and angle curves begins after the fault removal which corresponds to the decelerating area A_2. The fault clearing time determines the boundary at which the areas A_1 and A_2 are separated. Fault clearing is detectable by the opening position of the circuit-breakers which are tripped by the main protective relays. An alternative method is comparing P_m and P_e: right after the fault clearing, P_e becomes greater than P_m. This fact is used in the proposed algorithm for the detection of fault clearing time. It should be noted P_m is not measured independently, but assumed to be equal to the pre-disturbance value of P_e.

6.5   Simulation results

This section is devoted to evaluate the simulation results on three case studies ranging from small to large systems.

6.5.1   Case 1: single-machine system

The first study is a simple single-machine system shown in Figure 6.3 [27], where S2 is a synchronous generator and S1 and S3 are the fixed voltage source model of external grids. To create an OOS condition, a three-phase fault is applied on bus A and the fault clearing time is gradually increased until generator S2 becomes unstable. It was found that a 200-ms long fault starting at $t = 1$ s satisfies this goal.

The actual, ideal, and predicted power and angle curves are shown in Figure 6.4 for $F_r = 25$ Hz and $n_p = 2$. Figure 6.5 shows another example for $F_r = 10$ Hz and $n_p = 2$. In both figures, the predicted curves follow the ideal curves well. However, the accuracy of predicted curves in Figure 6.4 is more than Figure 6.5, because the former has a higher reporting rate than the latter. The ideal and actual

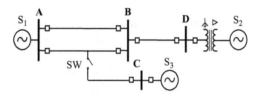

Figure 6.3   PSRC test system

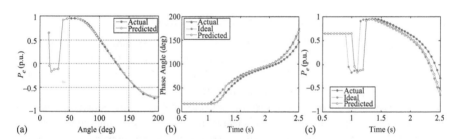

(a)   (b)   (c)

Figure 6.4   Simulation results for OOS following a three-phase fault at bus A with $F_r = 25$ Hz and $n_p = 2$; (a) angle δ, (b) electrical power, and (c) P–δ curve

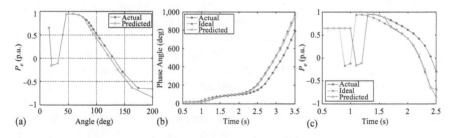

(a)   (b)   (c)

Figure 6.5   Simulation results for an OOS condition following a three-phase fault at bus A with $F_r = 10$ Hz and $n_p = 2$; (a) angle δ, (b) electrical power, and (c) P–δ curve

Table 6.2   Simulation results for the first case study

ARMA		F_r (Hz)	n_p	Predicted OOS instant (s)	Actual OOS instant (s)	Difference (ms)
p	q					
5	10	50	10	1.700	1.880	180
10	10	50	10	1.680	1.880	200
20	30	50	15	1.620	1.880	260
3	7	25	5	1.720	1.920	200
5	10	25	8	1.680	1.920	240
15	20	25	8	1.640	1.920	280
3	3	10	3	1.800	2.000	200
4	4	10	4	1.600	2.000	400

curves have completely overlapped with each other in Figures 6.4(c) and 6.5(c), since time is not represented in the P–δ plane and the power and angle curves treat as a parametric function of time.

Table 6.2 presents more simulation results with different ARMA model orders, reporting rates, and number of predicted samples. In this table, the actual and predicted OOS instants are obtained from EAC. It can be inferred that choosing appropriate values for F_r and n_p can expedite the prediction by hundreds of milliseconds.

6.5.2   Case 2: multi-machine two-area system

The second case study is conducted on the IEEE 39-bus system, which is a multi-machine system including 10 synchronous generators. This system is shown in Figure 6.6 and simulated in DIgSILENT Power Factory. OOS is simulated by applying a three-phase fault on F_1, located at the middle of lines 21 and 22. Figure 6.7(a–c) shows the electrical powers, rotor speeds, and rotor angles of the generators, respectively. The fault is applied at $t = 1$ s and cleared at $t = 1.2$ s by opening the lines 21 and 22. Upon the fault clearance, the system experiences severe power swings, during which generators G_6 and G_7 become unstable. Meanwhile, the coherency between G_6 and G_7 is clearly shown in Figure 6.7(b,c). After the outage of lines 21 and 22, there would be only one connection between the group of G_6 and G_7 and the rest of the grid via lines 23 and 24. This connection is not strong enough to damp the power swings, and G_6 and G_7 form a coherent group and oscillate against the rest of the grid.

To predict OOS, the EEAC is applied to the IEEE 39-bus system. The SMIB model of the system is built and then, the electrical power and the rotor angle of the SMIB model are predicted by the proposed ARMA/EKF approach. For instance, the critical and remaining clusters, as well as the actual and predicted P–δ curves for $F_r = 25$ Hz and $n_p = 3$ are illustrated in Figure 6.8. Results on more cases are also presented in Table 6.3. As shown, the proposed approach has successfully predicted the OOS condition in all cases. A proper time slot in the range of 160–300 ms is saved by the prediction, depending on the order of ARMA model, reporting frequency, and number of predicted samples. This example approves the efficiency of the proposed wide-area OOS protection in multi-machine power systems.

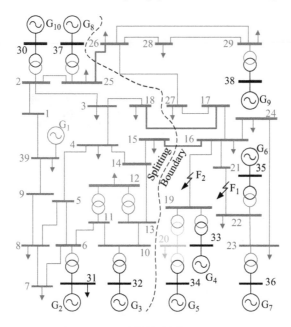

Figure 6.6   IEEE 39-bus system

Wide-area measurements may be polluted by noise contents. To investigate the effect of noisy measurements on the proposed method, white noises with various signal-to-noise ratios (SNRs) are added to the measured powers and angles. The applied EKF is updated to cancel out the noise effect by assigning random numbers to the elements of matrix **V** in (6.3) as $\mathbf{V}(m){\sim}N(0,0.2)$, where N stands for the normal probability distribution. For SNRs between 80 and 40 dB, there would be no considerable change in the obtained results. A more severe condition with SNR = 20 dB is investigated in Figure 6.9 where the actual, ideal, and predicted power curves are plotted altogether in the presence of noisy measurements. In spite of the heavy noise content, the red curve, representing the predicted curve, still follows the ideal curve in green with an acceptable tolerance.

Missing data are likely to be seen in wide-area measurements, due to some reasons such as communication failures. Performance of the proposed OOS prediction is shown in Figure 6.10 assuming that the WAMS data from PMUs 8, 9, and 10 are lost. Curves P1 and P2 show the predicted power when all PMU data are available and when the mentioned PMU data are lost, respectively. Meanwhile, curves P3 and P4 illustrate the ideal power curves in the presence of all PMU data and in the absence of three PMUs data, respectively. As shown, in spite of missing some data, the power curve P2 is well-estimated as P1.

To show the advantage of OOS prediction, an example of a remedial action which can be taken after the prediction is investigated in Figure 6.11. Here, a three-phase fault is applied on the middle of lines 16–19. In two scenarios, i.e. when the fault clearing time is 0.1 s and 0.3 s, the rotor angles are shown in Figure 6.11(a) and

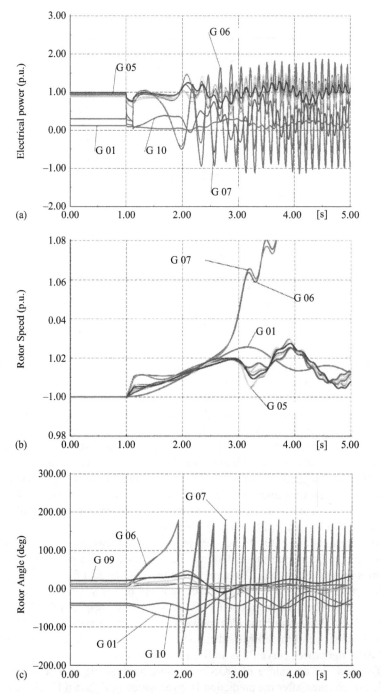

Figure 6.7   Simulation results for an OOS condition following a three-phase fault
on lines 21 and 22: (a) electrical powers, (b) rotor speeds, and
(c) rotor angles

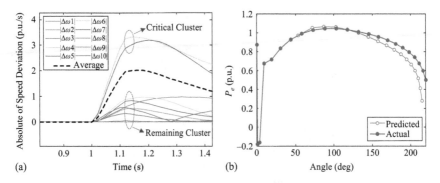

Figure 6.8 Generator clustering and EAC-based OOS detection: (a) critical and remaining clusters; (b) actual and predicted P–δ curves for $F_r = 25$ Hz and $n_p = 3$

Table 6.3 Evaluation of prediction accuracy

ARMA		F_r (Hz)	n_p	Predicted OOS instant (s)	Actual OOS instant (s)	Difference (ms)
p	q					
5	10	50	10	1.500	1.660	160
10	10	50	10	1.480	1.660	180
20	30	50	15	1.400	1.660	260
3	7	25	5	1.500	1.660	160
5	10	25	8	1.420	1.660	240
15	20	25	8	1.380	1.660	280
3	3	10	3	1.500	1.700	200
4	4	10	4	1.400	1.700	300

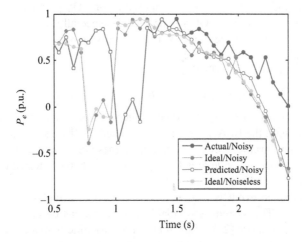

Figure 6.9 Ideal and predicted power curves in the presence of noisy measurements (with SNR = 20 dB)

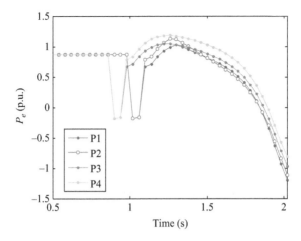

Figure 6.10   Investigation of communication failure effects on the predicted power curves

(b), respectively. For the fault clearing time of 0.3 s, some generators would experience OOS. One of the remedial actions which can be taken after the outage of lines 16–19 is to split the remaining network with respect to the coherency of generators. This action can be performed based on the pre-planned studies. Splitting the network from the boundary shown in Figure 6.6 needs to trip lines 25–26, 17–18, and 15–16.

To see what happens after the remedial action, the resulted rotor angles of the generators are depicted in Figure 6.11(c) and (d). If the remedial action is performed 400 ms after the outage of lines 16–19, all generators would remain stable, as shown in Figure 6.11(c). However, if the remedial action is applied after 600 ms, it would be useless and the system generators would be unable to recover their stability.

This case study demonstrates the importance of OOS prediction in preventing system instability. As soon as OOS is predicted (especially in the inter-area oscillation conditions), it would be more likely that the system is rescued from the threat of wide-spread instability. On the other hand, late remedial actions may be a postmortem solution and ineffective.

6.5.3   Case 3: multi-machine multi-area system

Large power systems often have hundreds of generation units and multiple areas. In such a large system, the EEAC can be used to detect inter-area loss of synchronism when the angular variations of the system machines are separable into two groups. In this part, the proposed wide-area OOS protection scheme is tested on a modified version of the IEEE 118-bus system [28].

The disturbance event is triggered by a three-phase fault occurring on lines 69–75 at $t = 1$ s. The system is dividable into five areas based on the structural decomposition procedure [29]. Thus, this is a 54-machine 5-area system, as shown in Figure 6.12. The machines' rotor angles within each area are depicted in Figure 6.13(a)–(e). Considering Figure 6.13(g) and based on the proposed

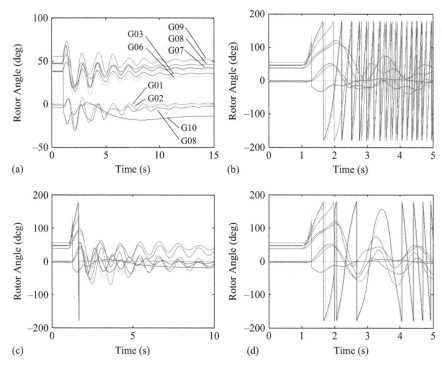

Figure 6.11   *Rotor angles variations when the faulted lines 16–19 is tripped after:*
(a) 0.1 s without remedial action; (b) 0.3 s without remedial action;
(c) 0.3 s and then the remedial action is applied after 0.4 s; (d) 0.3 s
and then the remedial action is applied after 0.6 s

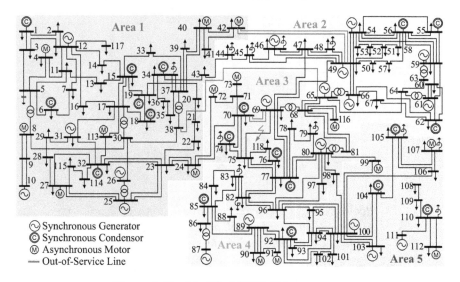

Figure 6.12   Modified IEEE 118-bus system

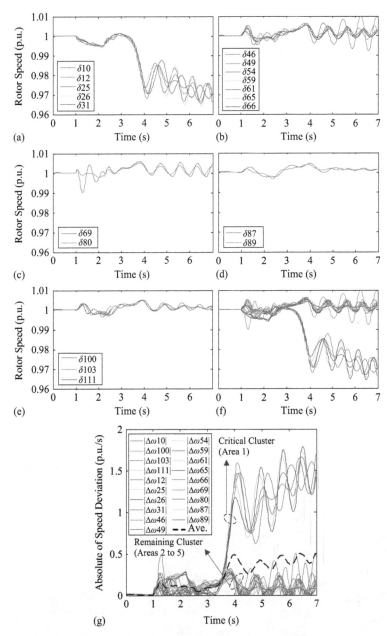

*Figure 6.13   Rotor speeds variations in: (a) Area 1; (b) Area 2; (c) Area 3;
(d) Area 4; (e) Area 5; (f) all areas; (g) absolutes of speed deviations
in critical and remaining clusters*

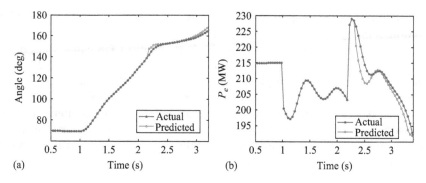

Figure 6.14   Simulation results for $F_r = 25$ Hz and $n_p = 3$ in the modified IEEE 118-bus system; (a) angle and (b) power

Table 6.4   Simulation results for the third case study

ARMA		F_r(Hz)	n_p	Predicted OOS instant (s)	Actual OOS instant (s)	Difference (ms)
p	q					
10	15	50	10	2.380	2.500	120
10	5	50	10	2.380	2.500	120
10	5	25	5	2.340	2.500	160
8	2	25	8	2.300	2.500	200
15	5	25	8	2.260	2.500	240
6	4	10	3	2.200	2.500	300

clustering method, area 1 is chosen in this event as the critical cluster. Meanwhile, the actual and predicted power and angles curves are illustrated in Figure 6.14 for $F_r = 25$ Hz and $n_p = 3$. Table 6.4 presents a summary of the obtained results which approve the capability of proposed method in prediction of OOS.

6.6   Conclusion

Wide-area measurements can enhance the efficiency and comprehensiveness of power system protection schemes. Thanks to having access to the current and previous states of the electrical parameters and variables coming from different parts of the grid, the traditional role of protection system as a corrective actor after the fault can be shifted to a predictive actor before the fault. This will help to save a golden time which can be used to apply effective remedial actions and prevent the aftermath of the major events endangering the security of power system. This chapter introduced a new predictive approach to wide-area out-of-step protection. A combination of ARMA modeling and EKF state estimation created a powerful prediction tool which was applied to the power and angles curves. Using the predictive curves, the transient stability of power system was assessed by EEAC in multi-machine systems to determine the risk of OOS.

It was shown that the proposed approach can predict OOS hundreds of milliseconds before it occurs. Of course, the time saved by the prediction depends on the reporting rate and number of predicted samples. One of the main features of the proposed method is that it does not require an exact knowledge of network configuration and detailed power system parameters. Indeed, it is a measurement-based approach and superior to the model-based methods. It was also demonstrated that the performance of proposed method is reliable when the measurements are noisy and some PMU data are lost.

References

[1]   Hashemi, S.M. and Sanaye-Pasand, M. A new predictive approach to wide-area out-of-step protection. *IEEE Trans. Ind. Informatics* 15(4), 1890–1898, 2019.
[2]   Xu, G., Vittal, V., Meklin, A., and Thalman, J.E. Controlled islanding demonstrations on the WECC system. *IEEE Trans. Power Syst.* 26(1), 334–343, 2011.
[3]   Zhang, S. and Zhang, Y. A novel out-of-step splitting protection based on the wide area information. *IEEE Trans. Smart Grid* 8(1), 41–51, 2017.
[4]   Alinezhad, B. and Kazemi Kargar, B. Out-of-step protection based on equal area criterion. *IEEE Trans. Power Syst.*, 32(2), 968–977, 2017.
[5]   Shrestha, B., Gokaraju, R., and Sachdev, M. Out-of-step protection using state-plane trajectories analysis. *IEEE Trans. Power Del.* 28(2), 1083–1093, 2013.
[6]   Amraee, T. and Ranjbar, S. Transient instability prediction using decision tree technique. *IEEE Trans. Power Syst.* 28(3), 3028–3037, 2013.
[7]   Li, M., Pal. A., Phadke, A.G., and Thorp, J.S. Transient stability prediction based on apparent impedance trajectory recorded by PMUs. *Int. J. Electr. Power Energy Syst.* 54, 498–504, 2014.
[8]   Wu, X., Xu, A., Zhao, J., Deng, H., and Xu, P. Review on transient stability prediction methods based on real time wide-area phasor measurements. In: *4th Int. Conf. on Electric Utility Deregulation and Restructuring and Power Technologies (DRPT)*, pp. 320–326. IEEE, China, 2011.
[9]   Salimian, M.R. and Aghamohammadi, M.R. A quantile regression-based approach for online probabilistic prediction of unstable groups of coherent generators in power systems. *IEEE Trans. Smart Grid* 9(4), 2288–2497, 2018.
[10] Shamisa, A., Majidi, B., and Patra, J.C. Sliding-window-based real-time model order reduction for stability prediction in smart grid. *IEEE Trans. Power Syst.* 34(1), 326–337, 2019.
[11] Das, S. and Panigrahi, B.K. Prediction and control of transient stability using system integrity protection schemes. *IET Gener. Transm. Distrib.* 13(8), 1247–1254, 2019.

[12] Zhu, L., Hill, D., and Lu, C. Hierarchical deep learning machine for power system online transient stability prediction. *IEEE Trans. Power Syst.* 35(3), 2399–2411, 2020.

[13] Azman, S.K., Isbeih, Y.J., El Moursi, M.S., and Elbassioni, K. A unified online deep learning prediction model for small signal and transient stability. *IEEE Trans. Power Syst.* 35(6), 4585–4598, 2020.

[14] Behdadnia, T., Yaslan, Y., and Genc, I. A new method of decision tree based transient stability assessment using hybrid simulation for real-time PMU measurements. *IET Gener. Transm. Distrib.* 15, 678–693, 2021.

[15] Tang, Y., Feng, L., Wang, Q., and Xu, Y. Hybrid method for power system transient stability prediction based on two-stage computing resources. *IET Gener. Transm. Distrib.* 12(8), 1697–1703, 2018.

[16] Kay, S.M. *Modern Spectral Estimation.* Prentice Hall, Hoboken, NJ, 1998.

[17] Brockwell, P.J. and Davis, R.A. *Introduction to Time Series and Forecasting*, 3rd ed., Springer, Switzerland, 2016.

[18] Haykin, S. *Adaptive Filter Theory.* Prentice Hall, Hoboken, NJ, 1996.

[19] Chehreghani Bozchalui, M. Application of wide area protection against power system oscillation. M.S. thesis, Dept. Electrical Eng., Univ. of Tehran, Iran, 2007.

[20] Montgomery, D.C., Jennings, C.L., and Kulahci, M. *Introduction to Time Series Analysis and Forecasting,* 2nd ed., Wiley, USA, 2015.

[21] IEEE Standard for Synchrophasor Measurements for Power Systems, IEEE Std C37.188. 1, 2011.

[22] Jonsson, M., Begovic, M., and Daalder, J. A new method suitable for real-time generator coherency determination. *IEEE Trans. Power Syst.* 19(3), 1473–1482, 2004.

[23] Joo, S., Liu, C., Jones, L.E., and Choe, J. Coherency and aggregation techniques incorporating rotor and voltage dynamics. *IEEE Trans. Power Syst.* 19(2), 1068–1075, 2004.

[24] Ariff, M.A.M. and Pal, B.C. Coherency identification in interconnected power system—an independent component analysis approach. *IEEE Trans. Power Syst.* 28(2), 1747–1755, 2013.

[25] Jiang, T., Jia, H., Yuan, H., Zhou, N., and Li, F. Projection pursuit: a general methodology of wide-area coherency detection in bulk power grid. *IEEE Trans. Power Syst.* 31(4), 2776–2786, 2016.

[26] Xue, Y. and Pavella, M. Critical-cluster identification in transient stability studies. *IEE Gener. Transm. Distrib.* 140(6), 481–489, 1993.

[27] IEEE PSRC: EMTP reference models for transmission line relay testing. http://www.pes-psrc.org.

[28] Center of Intelligent Systems & Networks (KIOS): Dynamic IEEE Test Systems. http://www.kios.ucy.ac.cy.

[29] Tuglie, E.D., Iannone, S.M., and Torelli, F. A coherency recognition based on structural decomposition procedure. *IEEE Trans. Power Syst.* 23(2), 555–563, 2008.

Chapter 7

Fault location algorithm for multi-terminal transmission system

Purushotham Reddy Chegireddy[1] and Ravikumar Bhimasingu[1]

This chapter proposes a multi-terminal fault location algorithm for transmission lines using synchrophasors, which have a lot of applications in the modern-day power system. Due to the presence of various sources and the presence of new lines, protection challenges faced by multi-terminal transmission lines are more compared to two-terminal transmission system. However, as the modern day power system is growing rapidly, the usage of multi-terminal transmission lines is preferable owing to their advantages possessed by them compared to two-terminal transmission system. So, to protect these lines against faults, protection of lines such as fault location and identification of the faulted section is an important task in the present electricity power grid. Due to various uncertainties such as line loading and weather conditions, the line parameters may change. So, a fault location algorithm without the use of line parameters makes the algorithm more precise, reliable and accurate. In this algorithm, voltage and currents available at all the source ends are utilized to estimate the fault location. The fault location has been estimated with variations of various line lengths, fault resistances, and fault type. An optimization problem has been created further and root mean error has been introduced to find the least value among all the sections and to identify the faulted section. To reduce the % error in fault location estimation, charging current compensation is introduced. It has been observed from the results that the maximum % error in the estimation of fault location has been reduced with the charging current compensation as compared to without charging current compensation. The performance of the algorithm has also been checked for variation in synchronization errors, mutual coupling, bad data, and communication failure.

[1]Department of Electrical Engineering, Indian Institute of Technology Hyderabad (IIT Hyderabad), India

Nomenclature

PMU　phasor measurement unit
KVL　Kirchhoff's voltage law
FFT　fast Fourier transform
GMD　geometrical mean distance
RME　root mean error
GPS　global positioning system

7.1　Introduction

Modern electricity grid has equipped with a lot of power transmission lines. Protection of such lines against various faults is a vital task else there are many chances of power system failure and there are chances of black-out due to cascade tripping of transmission lines as can be seen in [1–6]. Among the protection studies against faults, relay operation, relay coordination, fault detection, fault classification, fault location, fault identification, fault mitigation and quick restoration of the power grid are some of the tasks that should be carried to maintain the reliability of the power system [7–14]. Synchrophasors or phasor measurement units (PMUs), which provide accurate time stamping, have a lot of applications in the protection of these lines against such faults. The usage of PMUs in various protection aspects of transmission lines can be seen in [15–17].

Fault location algorithms are classified into three types: impedance-, traveling-, and intelligence-based techniques. Further these algorithms can be classified based on the requirement of data. In the literature, various authors proposed fault location algorithms by using the information from single-ended, double-ended, and multi-terminal data [18–60]. Single-ended fault location algorithms use voltage and current data available at one end to locate the fault. Double-ended fault location algorithms use the voltage and current data available at both ends of a transmission line. By providing taps to an already existing transmission line, a multi-terminal transmission line can be modeled, which gives a reliable power supply, as well as these lines, are economically viable. So, protection of these lines ensures that modern power grid operators can supply power to the consumers efficiently. Due to multiple sources which feed the fault, protection of such lines is a challenging task as compared to a two-terminal transmission system.

Fault location in multi-terminal transmission lines such as a teed or three-terminal transmission line or greater than three-terminal are proposed by various authors in the literature. Elsadd *et al.* [27] proposed an iterative-based fault location algorithm for two-terminal and three-terminal lines using unsynchronized voltage and current measurements. Dasilva *et al.* [28] proposed an algorithm for three-terminal lines by extracting phasors based on wavelets. Ahmadimanesh *et al.* [29] proposed a traveling wave-based three-terminal fault location algorithm using time–time transforms which does not use wavelets. Gaur *et al.* [30] proposed a three-terminal fault location algorithm using unsynchronized current measurements by using wavelets. However the sampling frequency is very high of the order 1 MHz. Lin *et al*. [31] proposed a three-terminal fault location technique using unsynchronized current and voltage data. However, the method is iterative. Mahamedi *et al*. [32] proposed a fault location algorithm for three-terminal transmission lines which does not require current data, and requires unsynchronized voltage data. However, symmetrical faults have not been considered. Mirzaei *et al*. [33] proposed a three-terminal fault location algorithm using deep learning method. However, such algorithm requires suitable training. Chen *et al*. [34] proposed a three-terminal fault location algorithm using synchronized data and by Taylor's series. Lin *et al*. [35] fault location algorithm is based on superimposed voltage and current phasors but line parameters are needed to evaluate fault location. The line parameters of a line change due to line loading and weather conditions. So, estimating the algorithm without using them makes the algorithm efficient. Abasi *et al*. [36] proposed a three-terminal fault location algorithm which does not require line parameters using synchronized voltage and current data. Davoudi *et al*. [37] proposed a three-terminal fault locator which is independent of line parameters. However, the algorithms require post fault data. Dian Lu *et al*. [38] proposed a fault location for three-terminal transmission lines which is independent of line parameters. However, the algorithm is iterative. The authors of [39–42] proposed non-iterative fault location algorithms for three-terminal lines without using line parameters. However, all these algorithms are limited to three-terminal transmission lines. So, developing a fault location algorithm beyond three terminals is also important in modern-day power systems.

In the literature, various authors proposed multi-terminal fault location algorithms which consist of greater than three terminals [43–60]. Ahmadimanesh *et al*. [43] proposed a multi-terminal fault location algorithm using S-transforms. Chaitanya *et al*. [44] proposed a traveling wave fault location algorithm in combination with a decision tree-based method for multi-terminal transmission lines where S-transforms have been used to calculate the arrival times. Ding *et al*. [45] proposed a traveling wave multi-terminal fault location algorithm which uses DWT and Clarke's transform in decomposing the current signals. Moravej *et al*. [46] proposed a fault location algorithm for multi-terminal transmission lines using Gabor Transforms whose computational burden is low. Ngu *et al*. [47] proposed a one-end multi-terminal fault location using impedance and traveling wave-based methods where the measurements at all the ends are taken independently. Yongli *et al*. [48] proposed a multi-terminal fault location using current traveling waves where FIMD and TEO are used in detecting the traveling wave arrival times. However, the sampling frequency is high for these algorithms where the sampling frequency is varied from 76.92 kHz to 2 MHz [43–48].

In impedance-based techniques, various authors proposed fault location algorithms using synchronized data or unsynchronized data. Funabashi *et al*. [49] proposed two methods for fault location on multi-terminal transmission lines using available current and voltage data. However, LLG fault and symmetrical faults have not been reported. Ghorbani *et al*. [50] proposed a multi-terminal fault location algorithm using the negative sequence method. However, symmetrical faults have not been reported. De Pereira *et al*. [51] proposed a multi-terminal fault location algorithm, where all faults have been reported, however, fault type has to be known as a pre-requisite in estimating the fault location. Brahma *et al.* [52] proposed a multi-terminal fault location algorithm using the synchronized voltage data available at all the terminals. The algorithm estimates source impedance and is iterative in estimating the fault location. Shoaib *et al.* [53] proposed a multi-terminal fault location algorithm using unsynchronized data which does not depend on the variation of source impedance. Jiang *et al.* [54] proposed a two-stage, non-iterative fault location algorithm for multi-terminal transmission lines using positive sequence components. Manassero *et al.* [55] proposed various mathematical models for various power system components for estimating fault location on multi-terminal transmission lines. Liu *et al.* [56] proposed a multi-terminal fault location algorithm by using the technique available for two-terminal transmission lines. Kai ping lien *et al.* [57] proposed a multi-terminal fault location using positive sequence phasors of voltage and current. Lee *et al.* [58] proposed a multi-terminal fault location algorithm for non-homogeneous transmission lines using synchrophasor data by extending the algorithms developed for two- and three-terminal networks. Abe *et al.* [59] proposed a multi-terminal fault location algorithm where the multi-terminal line is converted into a two-terminal transmission line which contains the point at which the fault has occurred. Cai *et al.* [60] proposed a multi-terminal fault location algorithm using synchrophasor data where after identifying the faulty branch, the problem of fault location in a multi-terminal is made into a problem of fault location of a two-terminal line. However, all these algorithms [49–60] require line parameters in estimating the fault location. So, in this chapter, a fault location estimation algorithm is proposed by using synchrophasor measurements available at all the terminals of the multi-terminal transmission network without the use of line parameters.

The main objectives of the proposed chapter are:

- Proposed multi-terminal fault location algorithm works for variation in fault type, fault location, and fault resistance.
- Proposed multi-terminal fault location algorithm is independent of line parameters.
- Proposed fault location algorithm is non-iterative and does not require zero sequence component data.
- Proposed fault location algorithm does not require pre-fault nor post-fault data in estimating the fault location.
- Proposed fault location algorithm requires lower sampling frequency of the order 4 kHz.

7.2 Derivation of fault location equations for multi-terminal networks

Figure 7.1 shows a multi-terminal network for deriving equations for a generalized multi-terminal network. For analysis purpose, we have considered a five-terminal (number of buses = N = 5) network, where it can be observed from Figure 7.2 that

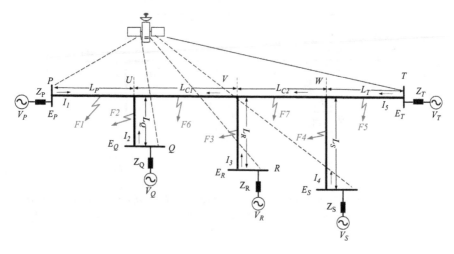

Figure 7.1 Single line diagram for generalizing equations for a multi-terminal network

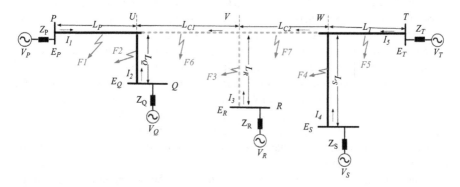

Figure 7.2 Splitting the multi-terminal network into main ends and intermediate ends

there are four main ends i.e. *P*, *Q*, *S*, and *T* and there are four ends *R*, *U*, *V*, and *W* which are intermediate ends which connect main ends such as *P* and *T*. If we further extend to six-terminal or seven-terminal, etc. also, the main ends will remain same, but the intermediate ends would increase further.

7.2.1 Derivation for asymmetrical fault location for fault on a main section which have sources at their ends-PU section

Assuming an asymmetrical fault say '*F1*' on *PU* section. Considering '*P*' as the main end, and naming it as the beginning end with 'subscript-*B*' and remaining ends *Q*, *R*, *S*, *T* as ending ends with 'subscript-*E*' (*E1*, *E2*, *E3*, *E4*). In between

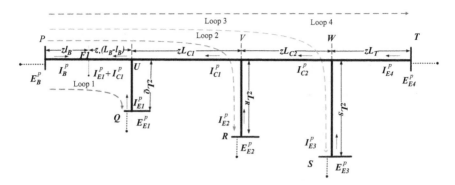

Figure 7.3   Positive sequence equivalent diagram of multi-terminal network for PU section with 2 interconnected branches

them, there are three more ends U, V, W. Indicating the main length L_P as L_B for *PU* section and line lengths of QU, RV, SW, TW as L_Q, L_R, L_S, L_T. The interconnected branches for the sections UV, VW here are indicated by L_{C1}, L_{C2}. Terminal voltages and currents flowing through the branches are indicated by letters 'E' and 'I'. 'z' is the positive/negative sequence impedance (in Ω/km) which is assumed the same. By splitting the network into positive ('superscript-p') and negative ('superscript-n') equivalent circuits as shown in Figures 7.3 and 7.4, it can be observed that there are 4 loops which contain the faulted location l_B, with 1 source 'P' (beginning) and 4 receiving (ending) ends ('Q, R, S, T'), and any 1 loop from positive equivalent circuit and any 1 loop from negative equivalent circuit will be used for the estimation of fault location. Hence, there would be 4 fault location estimation equations based on the number of loops for an asymmetrical fault on *PU* section.

The number of loop equations (NL) = (number of beginning ends)*(number of ending ends) = 1*4 = 4.

The number of asymmetrical fault location estimations for faults on PU section
$$= NL_{C_{NLj}} = \frac{NL!}{NLj!(NL-NLj)!} = {}^4C_1 = 4.$$

where 'NL' represents the number of loops formed while estimating the fault location, 'NLj' represents the number of loops used in one equivalent circuit. So, here '4(NL)' loops are formed while estimating the fault location on *PU* section and since at a time only '1(NLj)' loop from positive and negative equivalent circuit have been used, hence there would be 4 fault location estimations.

So if the fault occurred on *PU* section, fault location estimations can be done with the help of loops such as loop 1, loop 2, loop 3, and loop 4. Loop 1 touches the buses P and Q with base as P and end as Q. For this loop 1, the interconnected branches and their terms would not come. For loop 2, i.e., with base as P and end as R, the interconnected branch term with L_{C1} would come and for loop 3, loop 4, both the interconnected branch terms L_{C1}, L_{C2} would come. So, fault location equations can be written as per the loops.

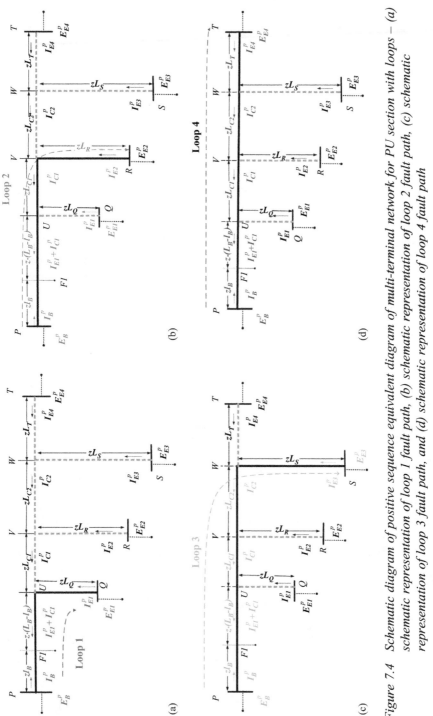

Figure 7.4 Schematic diagram of positive sequence equivalent diagram of multi-terminal network for PU section with loops – (a) schematic representation of loop 1 fault path, (b) schematic representation of loop 2 fault path, (c) schematic representation of loop 3 fault path, and (d) schematic representation of loop 4 fault path

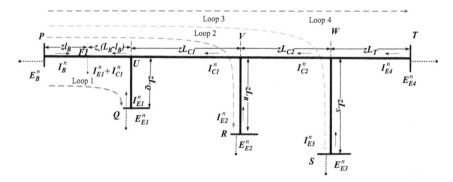

Figure 7.5   Negative sequence equivalent diagram of multi-terminal network for PU section with two interconnected branches

The schematic diagram of Figure 7.3 containing various loops in positive sequence equivalent diagram is mentioned separately in Figure 7.4. The thick line is associated with the corresponding loop taken.

Applying KVL to loop 1 for the positive sequence equivalent circuit of Figure 7.3, we can write,

$$-E_B^p + (zl_B)I_B^p - z(L_B - l_B)(I_{E1}^p + I_{C1}^p) - (zL_Q)I_{E1}^p + E_{E1}^p = 0 \qquad (7.1)$$

where the intermediate currents can be expressed in terms of source currents as, $I_{C2}^p = I_{E3}^p + I_{E4}^p$ and $I_{C1}^p = I_{E2}^p + I_{C2}^p = I_{E2}^p + I_{E3}^p + I_{E4}^p$.

Applying KVL to loop 1 for the negative sequence equivalent circuit of Figure 7.5, we can write,

$$-E_B^n + (zl_B)I_B^n - z(L_B - l_B)(I_{E1}^n + I_{C1}^n) - (zL_Q)I_{E1}^n + E_{E1}^n = 0 \qquad (7.2)$$

where $I_{C2}^n = I_{E3}^n + I_{E4}^n$ and $I_{C1}^n = I_{E2}^n + I_{C2}^n = I_{E2}^n + I_{E3}^n + I_{E4}^n$.

For simplicity, arranging the currents once again as per I_B as I_1, I_{E1} as I_2, I_{E2} as I_3, I_{E3} as I_4 and I_{E4} as I_5.

Rearranging the terms, the first fault location estimation l_B, say l_{B1} from loop 1 can be derived as (7.3).

$$l_{B1} = \frac{L_B\{E_{BE1}^p \displaystyle\sum_{\substack{i=1, \\ i \neq B}}^{N} I_i^n - E_{BE1}^n \sum_{\substack{i=1, \\ i \neq B}}^{N} I_i^p\} + L_Q\{E_{BE1}^p I_{E1}^n - E_{BE1}^n I_{E1}^p\}}{E_{BE1}^p \displaystyle\sum_{i=1}^{N} I_i^n - E_{BE1}^n \sum_{i=1}^{N} I_i^p} \qquad (7.3)$$

where $E_{BE1}^p = E_B^p - E_{E1}^p$, $E_{BE1}^n = E_B^n - E_{E1}^n$.

In the similar way by using loop 2, the second fault location estimation l_{B2} is derived as (7.4):

$$l_{B2} = \frac{L_B\{E_{BE2}^p \sum\limits_{\substack{i=1,\\i\neq B}}^{N} I_i^n - E_{BE2}^n \sum\limits_{\substack{i=1,\\i\neq B}}^{N} I_i^p\} + L_R\{E_{BE2}^p I_{E2}^n - E_{BE2}^n I_{E2}^p\} + L_{C1}\{E_{BE2}^p I_{C1}^n - E_{BE2}^n I_{C1}^p\}}{E_{BE2}^p \sum\limits_{i=1}^{N} I_i^n - E_{BE2}^n \sum\limits_{i=1}^{N} I_i^p}$$

(7.4)

By using loop 3, the third fault location estimation l_{B3} is derived as (7.5):

$$l_{B3} = \frac{L_B\{E_{BE3}^p \sum\limits_{\substack{i=1,\\i\neq B}}^{n} I_i^n - E_{BE3}^n \sum\limits_{\substack{i=1,\\i\neq B}}^{N} I_i^p\} + L_S\{E_{BE3}^p I_{E3}^n - E_{BE3}^n I_{E3}^p\} + L_{C1}\{E_{BE3}^p I_{C1}^n - E_{BE3}^n I_{C1}^p\}}{E_{BE3}^p \sum\limits_{i=1}^{N} I_i^n - E_{BE3}^n \sum\limits_{i=1}^{N} I_i^p}$$

$$+ \frac{L_{C2}\{E_{BE3}^p I_{C2}^n - E_{BE3}^n I_{C2}^p\}}{E_{BE3}^p \sum\limits_{i=1}^{N} I_i^n - E_{BE3}^n \sum\limits_{i=1}^{N} I_i^p}$$

(7.5)

By using loop 4, the fourth fault location estimation l_{B4} is derived as (7.6)

$$l_{B4} = \frac{L_B\{E_{BE4}^p \sum\limits_{\substack{i=1,\\i\neq B}}^{N} I_i^n - E_{BE4}^n \sum\limits_{\substack{i=1,\\i\neq B}}^{N} I_i^p\} + L_T\{E_{BE4}^p I_{E4}^n - E_{BE4}^n I_{E4}^p\} + L_{C1}\{(E_{BE4}^p I_{C1}^n - E_{BE4}^n I_{C1}^p\}}{E_{BE4}^p \sum\limits_{i=1}^{N} I_i^n - E_{BE4}^n \sum\limits_{i=1}^{N} I_i^p}$$

$$+ \frac{L_{C2}\{(E_{BE4}^p I_{C2}^n - E_{BE4}^n I_{C2}^p)\}}{E_{BE4}^p \sum\limits_{i=1}^{N} I_i^n - E_{BE4}^n \sum\limits_{i=1}^{N} I_i^p}$$

(7.6)

These fault location estimations, l_B^1, l_B^2, l_B^3, l_B^4 are complex values, where real part is taken into consideration. This consideration has been assumed throughout the study. The interconnected terms in these fault location estimations would not come if the loop does not go through the interconnected branches. Based on the loop equations, the terms would come accordingly. So, if we denote the receiving end line lengths L_Q, L_R, L_S, L_T with L_E in the above equations of (7.3)–(7.6), in general, we can express fault location equation for a N-terminal network in the form of:

$$I_B^i = \frac{L_B\{E_{BE}^p \sum\limits_{\substack{i=1,\\i\neq B}}^{N} I_i^n - E_{BE}^n \sum\limits_{\substack{i=1,\\i\neq B}}^{N} I_i^p\} + L_E\{E_{BE}^p I_E^n - E_{BE}^n I_E^p\} + L_{C1}\{E_{BE}^p I_{C1}^n - E_{BE}^n I_{C1}^p\}}{E_{BE}^p \sum\limits_{i=1}^{N} I_i^n - E_{BE}^n \sum\limits_{i=1}^{N} I_i^p}$$

$$\frac{+L_{C2}\{E_{BE}^p I_{C2}^n - E_{BE}^n I_{C2}^p\}\ldots\ldots\ldots + L_{C(N-3)}\{E_{BE}^p I_{C(N-3)}^n - E_{BE}^n I_{C(N-3)}^p\}}{E_{BE}^p \sum\limits_{i=1}^{N} I_i^n - E_{BE}^n \sum\limits_{i=1}^{N} I_i^p} \tag{7.7}$$

All these interconnected currents and voltages appearing though the branches, currents flowing through the branches should be decided based on the formation of loop equations. The same procedure is followed for main sections *QU, SW, TW* and the middle intermediate section having source i.e. *RV* section.

7.2.2   *Generalized asymmetrical fault location equations for sections which does not have source ends*

The same procedure which is discussed in Section 7.2.1 is followed for this analysis too. Since these sections does not have direct source ends, the loop equations are formed by taking some as sources and other as receiving ends based on the location of the fault. For fault *F6* on *UV* section, from Figures 7.6 and 7.7, '6(*NL*)' loops would come and '1(*NLj*)' loop from positive and negative equivalent circuit is to be used, with 2 beginning ends (*P,Q*) and 3 ending ends (*R,S,T*), hence there would be 6 fault location equations.

The number of asymmetrical fault location estimations for faults on *UV* section $=^{NL}C_{NLj}=^{6}C_1 = 6$.

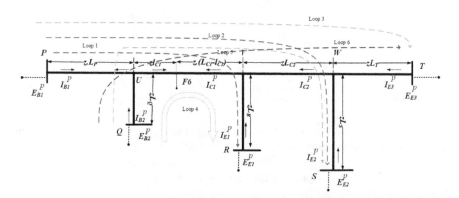

Figure 7.6   Positive sequence equivalent diagram for fault on UV section of a multi-terminal network with 2 interconnected branches

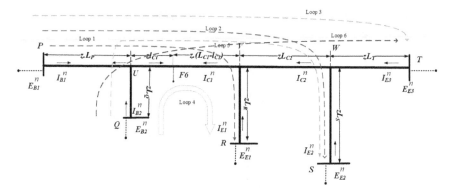

Figure 7.7   Negative sequence equivalent diagram for fault on UV section of a multi-terminal network with 2 interconnected branches

For a N-terminal generalized network, the fault location estimation will appear in form of:

$$
\begin{aligned}
l_C^i = & \frac{L_B\{E_{BE}^n I_B^p - E_{BE}^p I_B^n\} + L_E\{E_{BE}^p I_E^n - E_{BE}^n I_E^p\} + L_{C1}\{E_{BE}^p I_{C1}^n - E_{BE}^n I_{C1}^p\}}{E_{BE}^p \sum_{i=1}^{N} I_i^n - E_{BE}^n \sum_{i=1}^{N} I_i^p} \\
& \frac{+L_{C2}\{E_{BE}^p I_{C2}^n - E_{BE}^n I_{C2}^p\} + \ldots\ldots\ldots L_{C(N-3)}\{E_{BE}^p I_{C(N-3)}^n - E_{BE}^n I_{C(N-3)}^p\}}{E_{BE}^p \sum_{i=1}^{N} I_i^n - E_{BE}^n \sum_{i=1}^{N} I_i^p}
\end{aligned}
$$

$$(7.8)$$

All these interconnected currents and voltages appearing through the branches, currents flowing through the branches should be decided based on the formation of loop equations.

7.2.3   Generalized symmetrical fault location equations for faults which occurred at source ends

Since there is no negative sequence equivalent circuit involved, fault location estimation has to be derived by considering positive sequence components only. By using Figure 7.3, so, for a fault '$F1$' that occurred on PU section, fault location l_B is estimated by taking any two loops in positive sequence circuit only. As per this, '4 (NL)' loops would come, with 1 beginning end and 4 ending ends and '$2(NLj)$' loops at a time are to be used using positive equivalent circuit only. Hence there would be 6 fault location equations by using the loops *(PQ,PR), (PQ,PS), (PQ,PT), (PR,PS), (PR,PT), (PS,PT)* for fault on *PU* section.

The number of symmetrical fault location estimations for faults on *PU* section $=^{NL}C_{NLj} = {^4}C_2 = 6$.

For a N-terminal network, the generalized equation would appear in the form of:

$$l_B^i = \frac{L_B\{E_{E2E1}^p \sum_{\substack{i=1,\\i \neq B}}^{N} I_i^n\} + L_{C1}\{E_{x1x}^p I_{C1}^p\} + L_{C2}\{E_{y1y}^p I_{C2}^p\}\dots\dots + L_{C(N-3)}\{E_{z1z}^p I_{C(N-3)}^p\}}{E_{E2E1}^p \sum_{i=1}^{N} I_i^p}$$

$$\frac{+L_{E2}\{E_{BE1}^p I_{E2}^p\} - L_{E1}\{E_{BE2}^p I_{E1}^p\}}{E_{E2E1}^p \sum_{i=1}^{N} I_i^p}$$

(7.9)

The terms E_x^p, E_y^p, E_z^p, E_{x1}^p, E_{y1}^p, E_{z1}^p in the interconnected terms and the currents flowing through the interconnected branches would be based on the loop equations that have been considered.

7.2.4   Generalized symmetrical fault location equations for faults that occurred at sections which does not have sources

For a symmetrical fault on *UV* section as shown in Figure 7.6, the fault location estimation has been done by taking *P* and *Q* as base sources and *R*, *T*, and *S* as end receivers. So, there would be '6(*NL*)' loops. So, using any '2(*NLj*)' loops for fault location in the positive equivalent circuit, there would be 15 fault location estimation equations using the loops (*PR, PS*), (*PR, PT*), (*PR, QR*), (*PR, QS*), (*PR, QT*), (*PS, PT*), (*PS, QR*), (*PS, QS*), (*PS, QT*), (*PT, QR*), (*PT, QS*), (*PT, QT*), (*QR, QS*), (*QR, QT*), (*QS, QT*). The interconnected currents will come based on the currents flowing through the branches.

The number of symmetrical fault location estimations for faults on *UV* section $= {^{NL}}C_{NLj} = {^6}C_2 = 15$.

For a N-terminal network, the generalized equation will be in form of:

$$l_C^i = \frac{L_{B1}\{E_{x1x2}^p I_{B1}^p\} + L_{C1}\{(E_{x3x4}^p - E_{x5x6}^p)I_{C1}^p\} + L_{E2}\{E_{x7x8}^p I_{E2}^p\} - L_{E1}\{E_{x9x10}^p I_{E1}^p\}}{(E_{B1E1}^p - E_{B2E2}^p)\sum_{i=1}^{N} I_i^p}$$

$$\frac{-L_{B2}\{E_{x11x12}^p I_{B2}^p\} + L_{C2}\{E_{x13x14}^p - E_{x15x16}^p\}I_{C2}^p + \dots L_{C(N-3)}\{E_{xnxn1}^p - E_{xn2xn3}^p\}I_{C(N-3)}^p}{(E_{B1E1}^p - E_{B2E2}^p)\sum_{i=1}^{N} I_i^p}$$

(7.10)

The terms $E_{x1}, \ldots E_{xn3}$ would be based on the formation of loop equations by various sources, receivers and interconnecting ends.

7.3 Application of the proposed methodology on four-terminal network

As discussed in Section 7.2, for analysis purpose a four terminal ($N = 4$) network has been considered. Figure 7.8 shows the single-line diagram of the four-terminal homogeneous network that has been considered for fault location estimation. Table 7.1

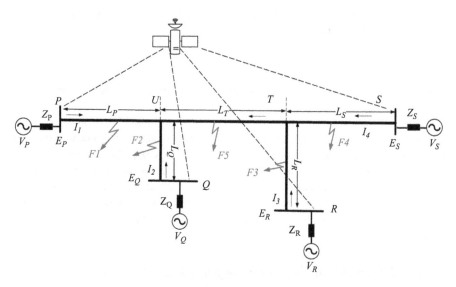

Figure 7.8   Single line diagram of considered four-terminal network

Table 7.1   Calculation of number of fault location estimations

Section	PU	QU	RT	ST	TU
Source end buses	P	Q	R	S	P & Q
Receiving end buses	Q, R, S	P, R, S	P, Q, S	P, Q, R	R & T
Number of loops in $(+)^{ve}/(-)^{ve}$ sequence circuit	3	3	3	3	4
Number of asymmetrical Fault location estimations	3 (I_P^1, I_P^2, I_P^3)	3 (I_Q^1, I_Q^2, I_Q^3)	3 (I_R^1, I_R^2, I_R^3)	3 (I_S^1, I_S^2, I_S^3)	4 $(I_T^1, I_T^2, I_T^3, I_T^4)$
Number of symmetrical Fault location estimations	3 (I_P^1, I_P^2, I_P^3)	3 (I_Q^1, I_Q^2, I_Q^3)	3 (I_R^1, I_R^2, I_R^3)	3 (I_S^1, I_S^2, I_S^3)	6 $(I_T^1, I_T^2, I_T^3, I_T^4, I_T^5, I_T^6)$

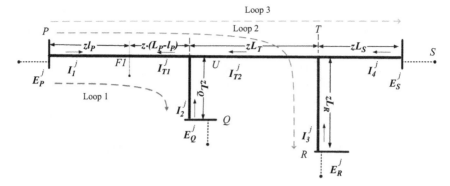

Figure 7.9   Positive/negative sequence equivalent diagram of considered four-terminal network

shows the number of loops and number of fault location estimations for various sections on this four-terminal network.

Figure 7.9 shows the equivalent circuits following a fault ($F1$) on PU section, 'j' can be 'p' (or) 'n' to represent the positive or negative sequence equivalent circuit of the four-terminal network. Here L_P, L_Q, L_R, L_S, L_T are the line lengths. (E_P^p, E_Q^p, E_R^p, E_S^p) and (E_P^p, E_Q^n, E_R^n, E_S^n) are the positive and negative sequence voltages, whereas (I_1^p, I_2^p, I_3^p, I_4^p) and (I_1^n, I_2^n, I_3^n, I_4^n) are the positive and negative sequence currents. These positive and negative sequence component data are obtained for the measurements (E_P, E_Q, E_R, E_S), (I_1, I_2, I_3, I_4) using PSCAD.

7.3.1   Asymmetrical fault location for faults on PU section

Following the faults on PU section, by using the generalized equation explained in Section 7.2.1, fault location estimations have been provided here for a four-terminal network of Figure 7.8. Also there would be formation of '3(NL)' loops and '1(NLj)' loop has to be taken from positive and negative equivalent circuits, there would be formation of 3 fault location estimations.

The number of asymmetrical fault location estimations for faults on PU section $= {}^{NL}C_{NLj} = {}^3C_1 = 3$.

If the same procedure as explained in Section 7.2.1 is applied here, the fault location estimation equations for faults on PU section are as given below:

$$l_P^1 = \frac{L_P\{E_{PQ}^p \sum_{i=2}^{4} I_i^n - E_{PQ}^n \sum_{i=2}^{4} I_i^p\} + L_Q\{E_{PQ}^p I_2^n - E_{PQ}^n I_2^p\}}{E_{PQ}^p \sum_{i=1}^{4} I_i^n - E_{PQ}^n \sum_{i=1}^{4} I_i^p} \tag{7.11}$$

$$l_P^2 = \frac{L_P\{E_{PR}^p \sum_{i=2}^{4} I_i^n - E_{PR}^n \sum_{i=2}^{4} I_i^p\} + L_T\{E_{PR}^p(I_3^n + I_4^n) - E_{PR}^n(I_3^p + I_4^p)\} + L_R\{E_{PR}^p I_3^n - E_{PR}^n I_3^p\}}{E_{PR}^p \sum_{i=1}^{4} I_i^n - E_{PR}^n \sum_{i=1}^{4} I_i^p}$$

$$(7.12)$$

$$l_P^3 = \frac{L_P\{E_{PS}^p \sum_{i=2}^{4} I_i^n - E_{PS}^n \sum_{i=2}^{4} I_i^p\} + L_T\{E_{PS}^p(I_3^n + I_4^n) - E_{PS}^n(I_3^p + I_4^p)\} + L_S\{E_{PS}^p I_4^n - E_{PS}^n I_4^p\}}{E_{PS}^p \sum_{i=1}^{4} I_i^n - E_{PS}^n \sum_{i=1}^{4} I_i^p}$$

$$(7.13)$$

To identify the exact fault location among the estimated fault locations, a optimization problem has been created, which is given as:

$$f_{PU}(x) = (x - x_1)^2 + (x - x_2)^2 + (x - x_3)^2 \tag{7.14}$$

Here the optimal value of 'x' will give the fault location x^* from the available fault location estimations x_1, x_2, x_3 which are here l_P^1, l_P^2, l_P^3 as mentioned above. To find the solution, differentiating the above function and equating it to zero, leads the optimal value of fault location, which is given as:

$$x^* = l_{P,opt} = \frac{l_P^1 + l_P^2 + l_P^3}{3} \tag{7.15}$$

Now from the obtained optimal fault location, the deviations are calculated from (7.16), for estimating the resultant value of fault location estimation l_P^R.

$$f_1 = |l_{P,opt} - l_P^1| \quad f_2 = |l_{P,opt} - l_P^2| \quad f_3 = |l_{P,opt} - l_P^3| \tag{7.16}$$

From the above if f_2 is low, then $l_P^2 (= l_P^R)$ is the resultant value of fault location for fault on PU section, similarly $l_Q^R, l_R^R, l_S^R, l_T^R$ are to be calculated in a similar way for the rest of the sections.

7.3.1.1   Fault section identification for faults on PU section

To identify the fault section identification, the root mean error value is calculated for all the sections. The section which gives the least value is considered as exact fault section.

The root mean error (RME) value for faults on *PU* section is given as:

$$RME_{PU} = \sqrt{\frac{f_{PU}(x)}{3}} \tag{7.17}$$

As an example, say an LG fault has occurred having a fault resistance of 0.1Ω on *PU* section at 10 km fault location. The estimated fault locations l_P^1, l_P^2, l_P^3, using (7.11)–(7.13) are respectively as 11.26 km, 11.12 km, and 10.73 km. To find the most relevant fault location, a function $f_{PU}(x)$ has been created as per (7.14). Differentiating and equating for zero leads to optimal value of $l_{P,opt}$ as per (7.15), which is calculated as 11.04 km. Now the deviation in fault location from optimum value has been estimated i.e., f_1, f_2, f_3 which came out to be 0.22 km, 0.08 km, 0.30 km. Clearly the value of f_2 is low, so the resultant fault location l_P^R is 11.12 km. Now as per (7.14), taking 'x' as '$l_{P,opt}$' and 'l_P^1' as x_1, 'l_P^2' as x_2, 'l_P^3' as x_3, the value of $f_{PU}(x)$ is calculated and later the value of RME_{PU} is calculated using (7.17) which came out to be 0.22 km.

The same procedure is followed for asymmetrical fault location estimations for QU, RT, ST, and TU sections.

7.3.2   *Symmetrical fault location for faults on PU section*

Using the generalized equation given in Section 7.2.3, following equations for fault on PU section can be derived for a four-terminal network of Figure 7.8. Here there would be '3(NL)' loops and '2(NLj)' loops are to be used from positive equivalent circuit only, hence there would be formation of 3 fault location estimations.

The number of symmetrical fault location estimations for faults on PU section $= {}^{NL}C_{NLj} = {}^3C_2 = 3$.

These fault location estimations are given for PU section as below:

$$l_P^1 = \frac{L_P\{E_{RQ}^p \sum_{i=2}^{4} I_i^p\} + L_T\{E_{PQ}^p(I_3^p + I_4^p)\} + L_R\{E_{PQ}^p I_3^p\} - L_Q\{E_{PR}^p I_2^p\}}{E_{RQ}^p \sum_{i=1}^{4} I_i^p}$$

(7.18)

$$l_P^2 = \frac{L_P\{E_{SQ}^p \sum_{i=2}^{4} I_i^p\} + L_T\{E_{PQ}^p(I_3^p + I_4^p)\} + L_S\{E_{PQ}^p I_4^p\} - L_Q\{E_{PS}^p I_2^p\}}{E_{SQ}^p \sum_{i=1}^{4} I_i^p}$$

(7.19)

$$l_P^3 = \frac{L_P\{E_{SR}^p \sum_{i=2}^{4} I_i^p\} + L_T\{E_{SR}^p(I_3^p + I_4^p)\} + L_S\{E_{PR}^p I_4^p\} - L_R\{E_{PS}^p I_3^p\}}{E_{SR}^p \sum_{i=1}^{4} I_i^p}$$

(7.20)

In a similar way, symmetrical fault location estimation equations have to be used for QU, RT, ST, and TU sections.

7.4   Charging current compensation

In the above derivation, the shunt capacitance of the transmission line is ignored. So, to reduce the effect of shunt capacitance, the shunt currents are to be evaluated and subtracted from I_1, I_2, I_3, I_4 for the estimation of fault location. For this purpose, the positive sequence capacitance is evaluated as from [61], assuming that the tower configuration is known. The charging current flowing through line capacitance at one end of PU section is given by:

$$I_{1,ch}^x = E_P^x \left(\frac{Y_p L_p}{2} \right) \tag{7.21}$$

where 'x' may take 'p' (or) 'n'. $Y_p = j B_P^p$, $B_P^p = 2\pi f C_P^p$ where C_P^p is the positive sequence capacitance which is calculated from tower configuration as per (7.22) as:

$$C_P^p = \frac{2\pi\varepsilon}{\ln\left(\frac{Dm}{r\sqrt{1+\left(\frac{Dm}{hm}\right)^2}} \right)} \tag{7.22}$$

where $\varepsilon = 8.854 * 10^{-12}$ F/m, r is the radius of the conductor in meters, D_m is the GMD in meters, and h_m is the mean conductor height over ground in meters. Similarly for QU, RT, ST sections also these currents are to be evaluated and for estimation of fault location these currents are subtracted from I_1, I_2, I_3, I_4.

7.5 Procedure for fault identification and fault location estimation for four-terminal network

Following a fault on any one of the section, following steps are to be followed for fault section identification and fault location estimation.

Step 1: All the synchrophasor voltage and current data are measured from all the (four) terminals.

Step 2: Calculate the sequence components of all the voltage and currents.

Step 3: Calculate the asymmetrical and symmetrical fault location estimations for all sections.

- Calculate the values of asymmetrical fault location estimations for all sections
  - l_P^1, l_P^2, l_P^3 for PU section using (7.11)–(7.13).
  - l_Q^1, l_Q^2, l_Q^3 for QU section.
  - l_R^1, l_R^2, l_R^3 for RT section.
  - l_S^1, l_S^2, l_S^3 for ST section.
  - l_T^1, l_T^2, l_T^3, l_T^4 for TU section.

- Calculate the values of symmetrical fault location estimations for all sections.
  - l_P^1, l_P^2, l_P^3 for PU section using (7.18)–(7.20).
  - l_Q^1, l_Q^2, l_Q^3 for QU section.
  - l_R^1, l_R^2, l_R^3 for RT section.

- ○ l_S^1, l_S^2, l_S^3 for ST section.
- ○ l_T^1, l_T^2, l_T^3, l_T^4, l_T^5, l_T^6 for TU section.

Step 4: Calculate the optimum value of fault locations for all the sections from the estimated fault locations.

- $l_{P,opt}$ for PU section using (7.15).
- $l_{Q,opt}$ for QU section.
- $l_{R,opt}$ for RT section.
- $l_{S,opt}$ for ST section.
- $l_{T,opt}$ for TU section.

Step 5: Calculate the deviations from optimal fault locations to estimated fault locations for all the sections.

- f_1, f_2, f_3 for PU section using (7.16).
- f_1, f_2, f_3 for QU section.
- f_1, f_2, f_3 for RT section.
- f_1, f_2, f_3 for ST section.
- f_1, f_2, f_3, f_4 for asymmetrical faults and $f_1, f_2, f_3, f_3, f_4, f_5, f_6$ for symmetrical faults on TU section.

Step 6: Calculate the resultant value of fault location i.e. the estimated fault locations whose deviations are less from optimal fault locations, using Step 5 for all the sections.

- l_P^R for PU section.
- l_Q^R for QU section.
- l_R^R for RT section.
- l_S^R for ST section.
- l_T^R for TU section.

Step 7: Calculate the RME by utilizing estimated fault locations and optimal fault locations for all the sections.

- RME_{PU} for PU section using (7.17).
- RME_{QU} for QU section.
- RME_{RT} for RT section.
- RME_{ST} for ST section.
- RME_{TU} for TU section.

Step 8: The section which gives the least value of *RME* is considered as fault section and estimate the resultant fault location.

The procedure for the fault location procedure is given in this flowchart as in Figure 7.10.

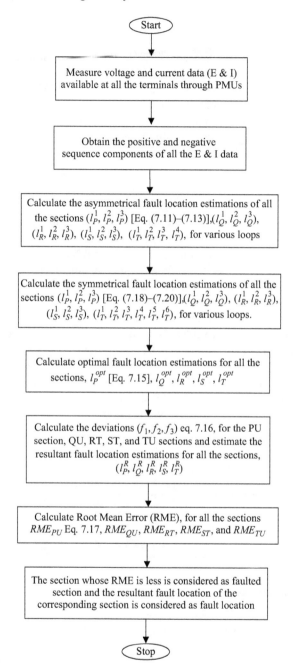

Figure 7.10   Flowchart for estimating the fault identification and fault location estimation for four-terminal network

7.6   Results

Tables 7.2 and 7.3 provide the 345 kV, 50 Hz network and transmission-line parameters that have been considered for simulation studies using PSCAD [62] and MATLAB®. Faults are created at 10%, 25%, 50%, 75%, and 90% of fault locations on all sections with 0.1Ω, 10Ω, and 100Ω as fault resistances. Using simulation studies, all the three phase voltages and currents are to be extracted at a rate of 50 frames/sec [63]. Thereby extract the fundamental positive and negative sequence phasor data. Faults have been applied at $t = 1$ s and second phasor is used for fault location estimation.

7.6.1   Fault section identification

Tables 7.4 and 7.5 provide the fault section identification for various faults at different fault locations with various fault resistances for without and with charging current compensation. Let us assume a LLG fault that has occurred on *PU* section with a fault resistance of 100Ω at a fault location of 90 km. From Figure 7.8, it can be observed that for fault on PU section, 3 fault location estimations (l_P^1, l_P^2, l_P^3) are calculated by using (7.11)–(7.13). The values of l_P^1, l_P^2, and l_P^3 came out to be 90.05 km, 90.47 km, and 90.39 km. After estimating these fault locations, an optimal value of fault location $l_{P,opt}$ is calculated using (7.15), which came out to be 90.30 km. Later the values of f_1, f_2, f_3 values are calculated using (7.16) which came out to be 0.25 km, 0.16 km, and 0.08 km. Clearly it can be observed that the value of f_3 is low, hence the resultant fault location l_P^R is 90.39 km. Similarly, these fault location estimations are to be carried for *QU*, *RT*, *ST*, and *TU* sections. After calculating all the various fault location estimations, the RME has to be calculated for all the sections as

Table 7.2   Network parameters

Source location	Terminal voltage (kV)	$Z_S^{+ve}(\Omega)$	$Z_S^0(\Omega)$
P	345 ∠ 30°	5.87 ∠ 87.73°	8.37∠ 77.69°
Q	345 ∠ 20°	6.23 ∠ 86.82°	7.87 ∠ 74.27°
R	345 ∠ 10 °	6.19 ∠ 88.16°	7.73 ∠ 76.65°
S	345 ∠ 0°	5.95 ∠ 87.90°	8.37 ∠ 79.03°

Table 7.3   Transmission line parameters

Parameter	Value	Parameter	Value
L_P	100 km	$R_1 = R_2$	0.034 (Ω/km)
L_Q	80 km	$L_1 = L_2$	1.33 (mH/km)
L_R	80 km	$C_1 = C_2$	8.74 (nF/km)
L_S	100 km	R_0	0.28 (Ω/km)
L_T	150 km	L_0	3.66 (mH/km)
		C_0	6 (nF/km)

Table 7.4 Fault section identification-without charging current compensation

Section	Fault type	R_F (Ω)	FL (km)	RME (km)					Section identified
				PU	QU	RT	ST	TU	
PU	LG	0.1	10	**0.22**	41.76	98.56	98.71	44.59	PU
	LL	10	50	**0.17**	23.11	84.59	84.67	24.71	PU
	LLG	100	90	**0.18**	4.49	72.95	73.00	4.98	PU
	LLLG	10	75	**0.91**	7.53	28.02	98.25	24.88	PU
QU	LG	100	8	34.06	**0.67**	92.01	92.27	35.60	QU
	LL	0.1	20	28.22	**0.57**	87.84	88.05	29.58	QU
	LLG	10	40	18.68	**0.46**	81.36	81.50	19.65	QU
	LLLG	100	72	5.44	**0.83**	49.25	133.88	7.36	QU
RT	LG	0.1	20	88.14	88.01	**0.25**	28.10	29.77	RT
	LL	100	60	75.68	75.62	**0.19**	9.21	9.90	RT
	LLG	10	72	72.42	72.37	**0.23**	3.46	3.91	RT
	LLLG	100	8	282.73	148.45	**0.64**	47.03	176.06	RT
ST	LG	10	10	98.68	98.57	41.79	**0.16**	44.58	ST
	LL	100	50	84.77	84.71	23.24	**0.11**	24.81	ST
	LLG	10	90	72.92	72.87	4.42	**0.27**	4.96	ST
	LLLG	0.1	75	88.22	30.72	7.77	**1.21**	23.12	ST
TU	LG	100	135	63.43	63.40	7.27	7.29	**0.03**	TU
	LL	0.1	112.5	52.88	52.86	17.82	17.83	**0.03**	TU
	LLG	10	15	7.31	7.31	63.37	63.41	**0.06**	TU
	LLLG	10	37.5	30.69	10.50	33.50	89.27	**0.37**	TU

Note: The minimum RME (Root Mean Error) value that identifies the faulted section among various sections is highlighted in bold values.

Table 7.5 Fault section identification-with charging current compensation

Section	Fault type	R_F (Ω)	FL (km)	RME (km)					Section identified
				PU	QU	RT	ST	TU	
PU	LG	0.1	10	**0.14**	41.96	98.69	98.78	44.72	PU
	LL	10	50	**0.13**	23.18	84.66	84.71	24.76	PU
	LLG	100	90	**0.14**	4.54	73.00	73.03	4.97	PU
	LLLG	10	75	**0.58**	7.43	27.88	98.05	24.85	PU
QU	LG	100	8	34.01	**0.42**	92.18	92.34	35.76	QU
	LL	0.1	20	28.19	**0.37**	87.98	88.11	29.71	QU
	LLG	10	40	18.68	**0.33**	81.46	81.55	19.74	QU
	LLLG	100	72	6.34	**0.43**	49.62	135.45	6.71	QU
RT	LG	0.1	20	88.17	88.10	**0.15**	28.11	29.84	RT
	LL	100	60	75.70	75.67	**0.14**	9.23	9.93	RT
	LLG	10	72	72.44	72.41	**0.20**	3.49	3.93	RT
	LLLG	100	8	281.52	148.47	**0.44**	47.11	181.22	RT
ST	LG	10	10	98.72	98.65	41.93	**0.09**	44.66	ST
	LL	100	50	84.80	84.76	23.30	**0.09**	24.85	ST
	LLG	10	90	72.94	72.91	4.45	**0.24**	4.96	ST
	LLLG	0.1	75	87.87	30.56	7.59	**0.75**	23.09	ST
	LG	100	135	63.45	63.43	7.25	7.27	**0.02**	TU
TU	LL	0.1	112.5	52.90	52.88	17.81	17.82	**0.02**	TU
	LLG	10	15	7.29	7.28	63.40	63.43	**0.04**	TU
	LLLG	10	37.5	30.84	10.49	33.29	88.96	**0.24**	TU

Note: The minimum RME (Root Mean Error) value that identifies the faulted section among various sections is highlighted in bold values.

calculated using (7.17) for *PU* section, to identify the exact fault section. From Table 7.4, the RME values for all the sections have come out to be 0.18 km, 4.49 km, 72.95 km, 73 km, and 4.98 km. Clearly, it can be observed that the root mean error value is less for *PU* section, which is highlighted in bold in the Table 7.4. Hence it can be said clearly that the fault has occurred on *PU* section at a fault location of 90.39 km. Similarly, the minimum RME value that identifies the faulted section among various sections is highlighted in bold values in Tables 7.4 and 7.5.

7.6.2   *Fault location estimation*

Figures 7.11–7.15 provide % error in fault location results for the case studies of without charging current compensation. Figures 7.16–7.20 provide the results for the % error in fault location results for the case studies with charging current

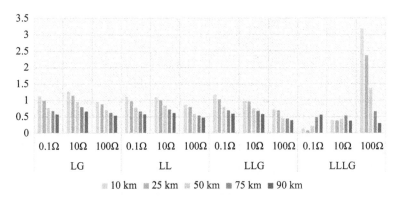

Figure 7.11   % error vs fault location without charging current compensation – PU section

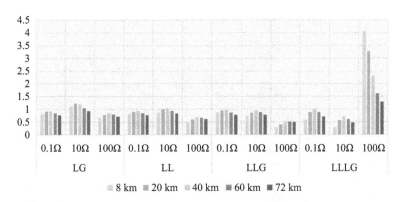

Figure 7.12   % error vs fault location without charging current compensation – QU section

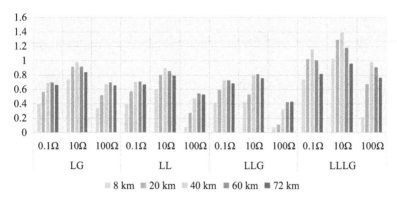

Figure 7.13   % error vs fault location without charging current compensation –
RT section

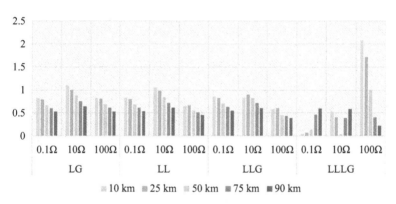

Figure 7.14   % error vs fault location without charging current compensation –
ST section

compensation. From these figures, it can be observed that the % error in fault location estimation is less than 1% for most of the cases considered. A total of 300 test cases have been studied from Figures 7.11–7.15 and it can be observed from these results that the maximum % error is found to be 4.07% and mean % error is found to be 0.69%. Similarly fault location results have been taken for those 300 test cases from Figures 7.16–7.20 by using charging current compensation to reduce the maximum % error, and it can be observed that the maximum and mean % errors have been reduced to 2.00% and 0.50% with charging current compensation. The % error in fault location for *PU* section is given as:

$$\%\text{error}\,(PU) = |\frac{l_P - l_P^R}{L_P}| * 100 \tag{7.23}$$

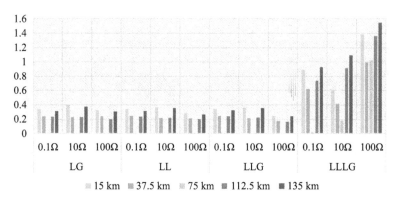

Figure 7.15   % error vs fault location without charging current compensation –
TU section

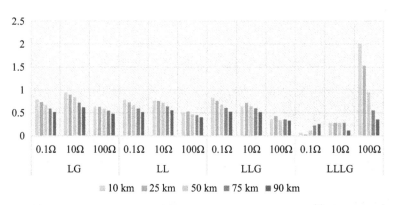

Figure 7.16   % error vs fault location with charging current compensation –
PU section

where l_P is the actual value of fault location and l_P^R is the resultant value of fault
location.

7.6.3   Synchronization error

GPS gives very precise time synchronization less than 1 µs. However, there is a
chance of happening slight synchronization errors. As per IEEE standard mentioned in [64], a phase error of 0.57°, corresponds to a time error of 31 µs for 50 Hz
system. So as per IEEE standard [64], in order to study the effect of synchronization error, here the synchrophasor data at the Q, R, and S terminals are assumed to
asynchronous, while P remains to be synchronized as given in Tables 7.6–7.10. In
these analysis, an LG fault with fault resistance of 0.1Ω has been applied to

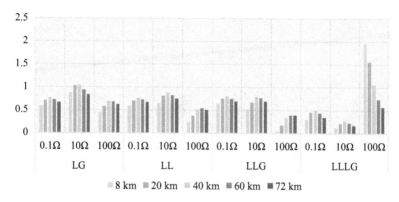

*Figure 7.17   % error vs fault location with charging current compensation –
QU section*

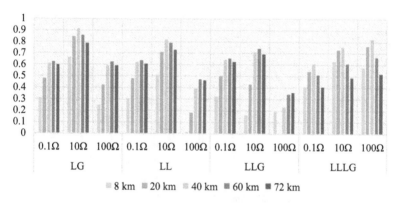

*Figure 7.18   % error vs fault location with charging current compensation –
RT section*

investigate the effect of synchronization errors. From these results, it can be
observed that lower synchronization errors such as 0.57°, cannot deteriorate the
fault location, whereas a delay of 4.5° that is a delay of 1 sample has caused some
increase in the % errors for many of the cases, where the maximum % error of 4.5%
without charging (w/o) current compensation and 4.6% with charging (w/c) current
compensation for a sample delay at *R* terminal for fault on *RT* section. However,
considering 2 samples delay i.e. 9° at *R* terminal and 4.5° each at *Q* and *S* terminals,
it can be observed that the % errors are too high for many of the cases and the
maximum % error is found to be 11.7% without charging current compensation and
12.2% with charging current compensation.

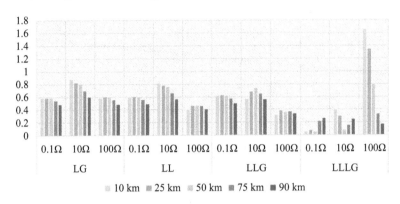

Figure 7.19   % error vs fault location with charging current compensation –
ST section

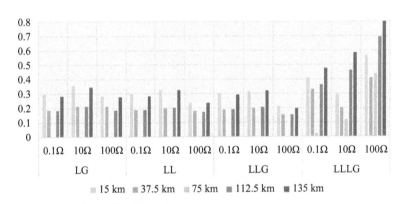

Figure 7.20   % error vs fault location with charging current compensation –
TU section

Table 7.6   Synchronization errors – % error vs fault location – PU section

$\delta_Q°$	$\delta_R°$	$\delta_S°$	10 km		25 km		50 km		75 km		90 km	
			w/o	w/c	w/o	w/c	w/o	w/c	w/o	w/c	w/o	w/c
–	–	0.57	1.09	0.76	0.96	0.71	0.61	0.48	0.55	0.46	0.47	0.41
4.5	4.5	–	2.01	2.49	1.66	2.06	1.17	1.46	0.75	0.93	0.12	0.13
	4.5	–	0.72	0.54	0.77	0.62	0.57	0.47	0.21	0.20	0.04	0.05
4.5	9	4.5	8.56	9.26	7.18	7.85	4.53	4.83	1.66	1.73	0.31	0.39

Table 7.7   Synchronization errors – % error vs fault location – QU section

$\delta_Q°$	$\delta_R°$	$\delta_S°$	8 km		20 km		40 km		60 km		72 km	
			w/o	w/c	w/o	w/c	w/o	w/c	w/o	w/c	w/o	w/c
–	–	0.57	0.35	0.10	0.48	0.27	0.58	0.41	0.57	0.44	0.54	0.43
4.5	4.5	–	2.26	1.55	2.03	1.45	1.63	1.23	1.26	0.99	1.05	0.85
	4.5	–	0.37	0.24	0.11	0.02	0.11	0.15	0.17	0.17	0.12	0.11
4.5	9	4.5	8.45	9.84	7.67	8.82	6.35	7.15	4.97	5.43	3.19	3.25

Table 7.8   Synchronization errors – % error vs fault location – RT section

$\delta_Q°$	$\delta_R°$	$\delta_S°$	8 km		20 km		40 km		60 km		72 km	
			w/o	w/c	w/o	w/c	w/o	w/c	w/o	w/c	w/o	w/c
–	–	0.57	0.40	0.31	0.56	0.47	0.68	0.60	0.69	0.62	0.65	0.59
4.5	4.5	–	0.41	0.14	0.60	0.35	0.74	0.53	0.74	0.58	0.70	0.57
	4.5	–	4.50	4.60	3.80	3.92	2.81	2.93	1.96	2.08	1.50	1.61
4.5	9	4.5	11.7	12.2	10.2	10.7	7.94	8.36	5.83	6.16	4.62	4.89

Table 7.9   Synchronization errors – % error vs fault location – ST section

$\delta_Q°$	$\delta_R°$	$\delta_S°$	10 km		25 km		50 km		75 km		90 km	
			w/o	w/c	w/o	w/c	w/o	w/c	w/o	w/c	w/o	w/c
–	–	0.57	0.28	0.14	0.40	0.29	0.50	0.41	0.48	0.42	0.43	0.38
4.5	4.5	–	2.74	2.56	2.53	2.39	2.10	2.03	0.83	0.83	0.49	0.45
	4.5	–	0.76	0.51	0.78	0.59	0.75	0.62	0.65	0.56	0.51	0.46
4.5	9	4.5	0.47	0.27	0.55	0.38	0.58	0.46	0.51	0.44	0.01	0.01

Table 7.10   Synchronization errors – % error vs fault location – TU section

$\delta_Q°$	$\delta_R°$	$\delta_S°$	15 km		37.5 km		75 km		112.5 km		135 km	
			w/o	w/c	w/o	w/c	w/o	w/c	w/o	w/c	w/o	w/c
–	–	0.57	0.30	0.27	0.18	0.17	0.004	0.004	0.22	0.16	0.30	0.28
4.5	4.5	–	0.30	0.28	0.26	0.22	0.12	0.11	0.19	0.17	0.27	0.25
	4.5	–	0.21	0.25	0.09	0.11	0.01	0.001	0.31	0.35	0.49	0.54
4.5	9	4.5	1.11	1.18	0.50	0.54	0.05	0.05	0.93	1.00	1.57	1.68

7.6.4   Effect of mutual coupling

The transmission line is modelled as frequency distributed transmission line model which is available in PSCAD, which also takes mutual coupling effect into consideration. However, here to see the effect, mutual coupling available between the phases is removed and simulation studies were studied by making the transmission line as 3-Ø RLC-Π section model with only self-elements. The studies were performed on PU section for faults at 10 km and 90 km. From Figure 7.21, it can be observed that the effect of mutual coupling is less comparable, however the maximum error which occurred for a symmetrical fault at 10 km with 100Ω is reduced if there is no mutual coupling as compared with mutual coupling effect.

7.6.5 Effect of bad data

The effect of bad data has been analyzed by taking measurement errors in voltage and current data for the faults on PU section as shown in Figure 7.22. A −10%

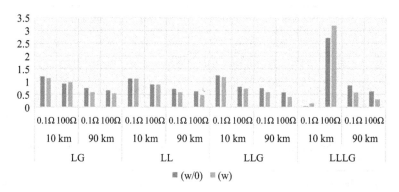

Figure 7.21   % error vs fault location without (w/o) and with (w) mutual coupling without charging current compensation – PU section

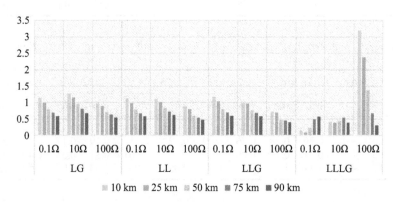

Figure 7.22   % error vs fault location with bad data – without charging current compensation – PU section

change in voltage data and +10% change in current data at all the terminals has been introduced. It has been observed that though with these variation there has not been any effect in the % error of fault location. This is because the fault location expressions which contain numerator and denominator parts which are in the form of (E1*I1)./(E2*I2) will get cancelled if we introduce the similar errors at all locations:

$$Ex: \frac{(E1*I1)}{(E2*I2)} = \frac{(0.9*E1 * 1.1*I1)}{(0.9*E2 * 1.1*I2)}$$
(7.24)

7.6.6 Communication failure

Communication failure and the error due to it mainly happens due to two factors which are latency and packet loss [65]. The PMUs which are located at different terminals may have various latencies, however PDC can align various PMUs data up to the availability of GPS. If there is no GPS, it may lead to synchronization errors, which has been discussed in Section 7.6.3.

Packet loss attributes to loss of data packets, which are measured at a rate of packets/sec (frames/sec). As the multi-terminal transmission network contains various sources of data, due to network congestion and transmission error issues, there might be delays in transferring the data from PMUs to PDC. In these certain cases, there would be loss of entire data from any of the PMU, and these are called packet losses. These packet losses are vulnerable in estimating the fault location. So, in such cases, the lost packets are to be replaced with some pseudo measurements [66]. To analyze the effect of packet loss, here it's assumed that due to communication failure the second packet or second phasor of faulted data has been lost at the Q terminal. Here, the analysis have performed for various fault conditions on PU section as in Figure 7.23, by replacing the original second phasor at Q terminal with other phasor such as prefault phasor data, first phasor of faulted data and third phasor of faulted data. It is observed that replacing with prefault data the % errors for both asymmetrical and symmetrical faults have been effected more. Utilizing first phasor of faulted data

Table 7.11 Effect of without (w/o) and with(w) mutual coupling without charging current compensation – % error vs fault location – PU section

Fault location (in km)	R_F (in Ω)	LG		LL		LLG		LLLG	
		w/o	w	w/o	w	w/o	w	w/o	w
10	0.1	1.19	1.12	1.11	1.12	1.24	1.16	0.02	0.13
10	100	0.91	0.96	0.88	0.87	0.77	0.72	2.70	3.19
90	0.1	0.73	0.57	0.70	0.57	0.73	0.58	0.85	0.55
90	100	0.64	0.53	0.61	0.46	0.56	0.39	0.61	0.30

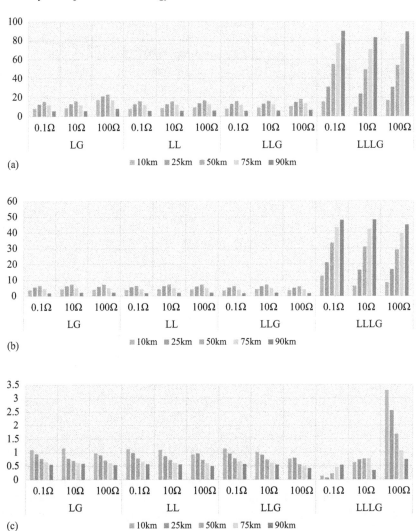

*Figure 7.23   % error vs fault location with pseudo measurement at Q terminal –
without charging current compensation – PU section: (a) using
pre-fault data as pseudo measurement, (b) using first phasor as
pseudo measurement, and (c) using third phasor as pseudo
measurement*

which is the sub-transient condition, it is observed that the % error have been
more for symmetrical faults. However, it is observed by utilizing third phasor
which is in transient condition, the % error have not effected much both for
asymmetrical and symmetrical faults.

7.7   Conclusion

This chapter presents a fault location algorithm for multi-terminal transmission lines using synchrophasor measurements. The proposed multi-terminal fault location algorithm uses positive and negative sequence components in estimating the fault location. The algorithm requires voltage and current data from all terminals in estimating the fault location. The algorithm is non-iterative and requires only faulted data in estimating the fault location. The fault location algorithm proposes a root mean error estimation method in order to identify the faulted section and thereby optimal and resultant values of fault location has been calculated. The algorithm also proposes a charging current compensation method to reduce the maximum % error in estimating the fault location. The mean % errors in estimating the fault location with and without charging current compensation method are given by 0.50% and 0.69%. The proposed fault location algorithm also shows effect of fault type, fault resistance, fault location, and effect of synchronization error. It has been observed that the fault location algorithm has not been effected with the variation of small synchronization errors. However, it is not true if the synchronization errors are increased. It is also observed that the % error has not much varied too much without mutual coupling effect. In case of similar measurement errors at all terminals, there is no effect on % error in estimation of fault location. The packet loss effect have been analyzed by replacing the lost packet with pseudo measurement data.

7.8   Appendix

Apart from PU section whose equations are given in Section 7.3, the equations belonging to four-terminal network for rest of the sections are presented here.

7.8.1   *Asymmetrical fault location for faults on QU section*
7.8.1.1   Fault location estimations

$$l_Q^1 = \frac{L_Q\{E_{QP}^p \sum_{\substack{i=1,\\i\neq2}}^{4} I_i^n - E_{QP}^n \sum_{\substack{i=1,\\i\neq2}}^{4} I_i^n\} + L_P\{E_{QP}^p I_1^n - E_{QP}^n I_1^p\}}{E_{QP}^p \sum_{i=1}^{4} I_i^n - E_{QP}^n \sum_{i=1}^{4} I_i^p} \qquad (7.25)$$

$$l_Q^2 = \frac{L_Q\{E_{QR}^p \sum_{\substack{i=1,\\i\neq2}}^{4} I_i^n - E_{QR}^n \sum_{\substack{i=1,\\i\neq2}}^{4} I_i^p\} + L_T\{E_{QR}^p(I_3^n + I_4^n) - E_{QR}^n(I_3^p + I_4^p)\} + L_R\{E_{QR}^p I_3^n - E_{QR}^n I_3^p\}}{E_{QR}^p \sum_{i=1}^{4} I_i^n - E_{QR}^n \sum_{i=1}^{4} I_i^p}$$

$$(7.26)$$

$$l_Q^3 = \frac{L_Q\{E_{QS}^p \sum\limits_{\substack{i=1, \\ i \neq 2}}^{4} I_i^n - E_{QS}^n \sum\limits_{\substack{i=1, \\ i \neq 2}}^{4} I_i^p\} + L_T\{E_{QS}^p(I_3^n + I_4^n) - E_{QS}^n(I_3^p + I_4^p)\} + L_S\{E_{QS}^p I_4^n - E_{QS}^n I_4^p\}}{E_{QS}^p \sum\limits_{i=1}^{4} I_i^n - E_{QS}^n \sum\limits_{i=1}^{4} I_i^p}$$

$$(7.27)$$

7.8.1.2   Optimal fault location estimation

$$l_{Q,opt} = \frac{l_Q^1 + l_Q^2 + l_Q^3}{3} \tag{7.28}$$

7.8.1.3   Deviations and estimation of resultant fault location estimation

$$f_1 = |l_{Q,opt} - l_Q^1| \quad f_2 = |l_{Q,opt} - l_Q^2| \quad f_3 = |l_{Q,opt} - l_Q^3| \tag{7.29}$$

From the above if f_2 is low, then $l_Q^2 \left(= l_Q^R\right)$ is the resultant value of fault location for fault on QU section.

7.8.1.4   Root mean error

$$f_{QU}(x) = (f_1)^2 + (f_2)^2 + (f_3)^2 \tag{7.30}$$

$$RME_{QU} = \sqrt{\frac{f_{QU}(x)}{3}} \tag{7.31}$$

7.8.2   Asymmetrical fault location for faults on RT section
7.8.2.1   Fault location estimations

$$l_R^1 = \frac{L_R\{E_{RS}^p \sum\limits_{\substack{i=1, \\ i \neq 3}}^{4} I_i^n - E_{RS}^n \sum\limits_{\substack{i=1, \\ i \neq 3}}^{4} I_i^p\} + L_S\{E_{RS}^p I_4^n - E_{RS}^n I_4^p\}}{E_{RS}^p \sum\limits_{i=1}^{4} I_i^n - E_{RS}^n \sum\limits_{i=1}^{4} I_i^p}$$

$$(7.32)$$

$$l_R^2 = \frac{L_R\{E_{RQ}^p \sum\limits_{\substack{i=1, \\ i \neq 3}}^{4} I_i^n - E_{RQ}^n \sum\limits_{\substack{i=1, \\ i \neq 3}}^{4} I_i^p\} + L_T\{E_{RQ}^p(I_1^n + I_2^n) - E_{RQ}^n(I_1^p + I_2^p)\} + L_Q\{E_{RQ}^p I_2^n - E_{RQ}^n I_2^p\}}{E_{RQ}^p \sum\limits_{i=1}^{4} I_i^n - E_{RQ}^n \sum\limits_{i=1}^{4} I_i^p}$$

$$(7.33)$$

$$l_R^3 = \frac{L_R\{E_{RP}^p \sum\limits_{\substack{i=1,\\i\neq 3}}^{4} I_i^n - E_{RP}^n \sum\limits_{\substack{i=1,\\i\neq 3}}^{4} I_i^p\} + L_T\{E_{RP}^p(I_1^n + I_2^n) - E_{RP}^n(I_1^p + I_2^p)\} + L_P\{E_{RP}^p I_1^n - E_{RP}^n I_1^p\}}{E_{RP}^p \sum\limits_{i=1}^{4} I_i^n - E_{RP}^n \sum\limits_{i=1}^{4} I_i^p}$$

(7.34)

7.8.2.2 Optimal fault location estimation

$$l_{R,opt} = \frac{l_R^1 + l_R^2 + l_R^3}{3}$$

(7.35)

7.8.2.3 Deviations and estimation of resultant fault location estimation

$$f_1 = |l_{R,opt} - l_R^1| \quad f_2 = |l_{R,opt} - l_R^2| \quad f_3 = |l_{R,opt} - l_R^3|$$

(7.36)

From the above if f_2 is low, then $l_R^2 (= l_R^R)$ is the resultant value of fault location for fault on RT section.

7.8.2.4 Root mean error

$$f_{RT}(x) = (f_1)^2 + (f_2)^2 + (f_3)^2$$

(7.37)

$$RME_{RT} = \sqrt{\frac{f_{RT}(x)}{3}}$$

(7.38)

7.8.3 Asymmetrical fault location for faults on ST section
7.8.3.1 Fault location estimations

$$l_S^1 = \frac{L_S\{E_{SR}^p \sum\limits_{\substack{i=1,\\i\neq 4}}^{4} I_i^n - E_{SR}^n \sum\limits_{\substack{i=1,\\i\neq 4}}^{4} I_i^p\} + L_R\{E_{SR}^p I_3^n - E_{SR}^n I_3^p\}}{E_{SR}^p \sum\limits_{i=1}^{4} I_i^n - E_{SR}^n \sum\limits_{i=1}^{4} I_i^p}$$

(7.39)

$$l_S^2 = \frac{L_S\{E_{SQ}^p \sum\limits_{\substack{i=1,\\i\neq 4}}^{4} I_i^n - E_{SQ}^n \sum\limits_{\substack{i=1,\\i\neq 4}}^{4} I_i^p\} + L_T\{E_{SQ}^p(I_1^n + I_2^n) - E_{SQ}^n(I_1^p + I_2^p)\} + L_Q\{E_{SQ}^p I_2^n - E_{SQ}^n I_2^p\}}{E_{SQ}^p \sum\limits_{i=1}^{4} I_i^n - E_{SQ}^n \sum\limits_{i=1}^{4} I_i^p}$$

(7.40)

$$l_S^3 = \frac{L_S\{E_{SP}^p \sum\limits_{\substack{i=1,\\i\neq 4}}^{4} I_i^n - E_{SP}^n \sum\limits_{\substack{i=1,\\i\neq 4}}^{4} I_i^p\} + L_T\{E_{SP}^p(I_1^n + I_2^n) - E_{SP}^n(I_1^p + I_2^p)\} + L_P\{E_{SP}^p I_1^n - E_{SP}^n I_1^p\}}{E_{SP}^p \sum\limits_{i=1}^{4} I_i^n - E_{SP}^n \sum\limits_{i=1}^{4} I_i^p}$$

$$(7.41)$$

7.8.3.2   Optimal fault location estimation

$$l_{S,opt} = \frac{l_S^1 + l_S^2 + l_S^3}{3} \tag{7.42}$$

7.8.3.3   Deviations and estimation of resultant fault location estimation

$$f_1 = |l_{S,opt} - l_S^1| \quad f_2 = |l_{S,opt} - l_S^2| \quad f_3 = |l_{S,opt} - l_S^3| \tag{7.43}$$

From the above if f_2 is low, then $l_S^2(= l_S^R)$ is the resultant value of fault location for fault on ST section.

7.8.3.4   Root mean error

$$f_{ST}(x) = (f_1)^2 + (f_2)^2 + (f_3)^2 \tag{7.44}$$

$$RME_{ST} = \sqrt{\frac{f_{ST}(x)}{3}} \tag{7.45}$$

7.8.4   *Asymmetrical fault location for faults on TU section*
7.8.4.1   Fault location estimations

$$l_T^1 = \frac{L_T\{E_{QR}^p(I_3^n + I_4^n) - E_{QR}^n(I_3^p + I_4^p)\} + L_Q\{E_{QR}^n I_2^p - E_{QR}^p I_2^n\} + L_R\{E_{QR}^p I_3^n - E_{QR}^n I_3^p\}}{E_{QR}^p \sum\limits_{i=1}^{4} I_i^n - E_{QR}^n \sum\limits_{i=1}^{4} I_i^p}$$

$$(7.46)$$

$$l_T^2 = \frac{L_T\{E_{QS}^p(I_3^n + I_4^n) - E_{QS}^n(I_3^p + I_4^p)\} + L_Q\{E_{QS}^n I_2^p - E_{QS}^p I_2^n\} + L_S\{E_{QS}^p I_4^n - E_{QS}^n I_4^p\}}{E_{QS}^p \sum\limits_{i=1}^{4} I_i^n - E_{QS}^n \sum\limits_{i=1}^{4} I_i^p}$$

$$(7.47)$$

$$l_T^3 = \frac{L_T\{E_{PR}^p(I_3^n + I_4^n) - E_{PR}^n(I_3^p + I_4^p)\} + L_P\{E_{PR}^n I_1^p - E_{PR}^p I_1^n\} + L_R\{E_{PR}^p I_3^n - E_{PR}^n I_3^p\}}{E_{PR}^p \sum\limits_{i=1}^{4} I_i^n - E_{PR}^n \sum\limits_{i=1}^{4} I_i^p}$$

$$(7.48)$$

$$l_T^4 = \frac{L_T\{E_{PS}^p(I_3^n + I_4^n) - E_{PS}^n(I_3^p + I_4^p)\} + L_P\{E_{PS}^n I_1^p - E_{PS}^p I_1^n\} + L_S\{E_{PS}^p I_4^n - E_{PS}^n I_4^p\}}{E_{PS}^p \sum_{i=1}^{4} I_i^n - E_{PS}^n \sum_{i=1}^{4} I_i^p}$$

$$(7.49)$$

7.8.4.2 Optimal fault location estimation

$$l_{T,opt} = \frac{l_T^1 + l_T^2 + l_T^3 + l_T^4}{4} \tag{7.50}$$

7.8.4.3 Deviations and estimation of resultant fault location estimation

$$f_1 = |l_{T,opt} - l_T^1| \quad f_2 = |l_{T,opt} - l_T^2| \tag{7.51}$$

$$f_3 = |l_{T,opt} - l_T^3| \quad f_4 = |l_{T,opt} - l_T^4| \tag{7.52}$$

From the above if f_2 is low, then $l_T^2 (= l_T^R)$ is the resultant value of fault location for fault on TU section.

7.8.4.4 Root mean error

$$f_{TU}(x) = (f_1)^2 + (f_2)^2 + (f_3)^2 + (f_4)^2 \tag{7.53}$$

$$RME_{TU} = \sqrt{\frac{f_{TU}(x)}{3}} \tag{7.54}$$

7.8.5 Symmetrical fault location for faults on QU section
7.8.5.1 Fault location estimations

$$l_Q^1 = \frac{L_Q\{E_{RP}^p \sum_{\substack{i=1,\\i\neq 2}}^{4} I_i^p\} + L_T\{E_{QP}^p(I_3^p + I_4^p)\} + L_R\{E_{QP}^p I_3^p\} - L_P\{E_{QR}^p I_1^p\}}{E_{RP}^p \sum_{i=1}^{4} I_i^p}$$

$$(7.55)$$

$$l_Q^2 = \frac{L_Q\{E_{SP}^p \sum_{\substack{i=1,\\i\neq 2}}^{4} I_i^p\} + L_T\{E_{QP}^p(I_3^p + I_4^p)\} + L_S\{E_{QP}^p I_4^p\} - L_P\{E_{QS}^p I_1^p\}}{E_{SP}^p \sum_{i=1}^{4} I_i^p}$$

$$(7.56)$$

$$l_Q^3 = \frac{L_Q\{E_{SR}^p \sum\limits_{\substack{i=1, \\ i \neq 2}}^{4} I_i^p\} + L_T\{E_{SR}^p(I_3^p + I_4^p)\} + L_R\{E_{SQ}^p I_3^p\} - L_S\{E_{RQ}^p I_4^p\}}{E_{SR}^p \sum\limits_{i=1}^{4} I_i^p}$$

(7.57)

7.8.5.2   Optimal fault location estimation

$$l_{Q,opt} = \frac{l_Q^1 + l_Q^2 + l_Q^3}{3}$$

(7.58)

7.8.5.3   Deviations and estimation of resultant fault location estimation

$$f_1 = |l_{Q,opt} - l_Q^1| \quad f_2 = |l_{Q,opt} - l_Q^2| \quad f_3 = |l_{Q,opt} - l_Q^3|$$

(7.59)

From the above if f_2 is low, then $l_Q^2 \left(= l_Q^R\right)$ is the resultant value of fault location for fault on QU section.

7.8.5.4   Root mean error

$$f_{QU}(x) = (f_1)^2 + (f_2)^2 + (f_3)^2$$

(7.60)

$$RME_{QU} = \sqrt{\frac{f_{QU}(x)}{3}}$$

(7.61)

7.8.6   *Symmetrical fault location for faults on RT section*
7.8.6.1   Fault location estimations

$$l_R^1 = \frac{L_R\{E_{QS}^p \sum\limits_{\substack{i=1, \\ i \neq 3}}^{4} I_i^p\} + L_T\{E_{RS}^p(I_1^p + I_2^p)\} + L_Q\{E_{RS}^p I_2^p\} - L_S\{E_{RQ}^p I_4^p\}}{E_{QS}^p \sum\limits_{i=1}^{4} I_i^p}$$

(7.62)

$$l_R^2 = \frac{L_R\{E_{PS}^p \sum\limits_{\substack{i=1, \\ i \neq 3}}^{4} I_i^p\} + L_T\{E_{RS}^p(I_1^p + I_2^p)\} + L_P\{E_{RS}^p I_1^p\} - L_S\{E_{RP}^p I_4^p\}}{E_{PS}^p \sum\limits_{i=1}^{4} I_i^p}$$

(7.63)

$$l_R^3 = \frac{L_R\{E_{PQ}^p \sum\limits_{\substack{i=1,\\i\neq 3}}^{4} I_i^p\} + L_T\{E_{PQ}^p(I_1^p + I_2^p)\} + L_P\{E_{RQ}^p I_1^p\} - L_Q\{E_{RP}^p I_2^p\}}{E_{PQ}^p \sum\limits_{i=1}^{4} I_i^p}$$

(7.64)

7.8.6.2 Optimal fault location estimation

$$l_{R,opt} = \frac{l_R^1 + l_R^2 + l_R^3}{3}$$

(7.65)

7.8.6.3 Deviations and estimation of resultant fault location estimation

$$f_1 = |l_{R,opt} - l_R^1| \quad f_2 = |l_{R,opt} - l_R^2| \quad f_3 = |l_{R,opt} - l_R^3|$$

(7.66)

From the above if f_2 is low, then $l_R^2 (= l_R^R)$ is the resultant value of fault location for fault on RT section.

7.8.6.4 Root mean error

$$f_{RT}(x) = (f_1)^2 + (f_2)^2 + (f_3)^2$$

(7.67)

$$RME_{RT} = \sqrt{\frac{f_{RT}(x)}{3}}$$

(7.68)

7.8.7 Symmetrical fault location for faults on ST section
7.8.7.1 Fault location estimations

$$l_S^1 = \frac{L_S\{E_{QR}^p \sum\limits_{\substack{i=1,\\i\neq 4}}^{4} I_i^p\} + L_T\{E_{SR}^p(I_1^p + I_2^p)\} + L_Q\{E_{SR}^p I_2^p\} - L_R\{E_{SQ}^p I_3^p\}}{E_{QR}^p \sum\limits_{i=1}^{4} I_i^p}$$

(7.69)

$$l_S^2 = \frac{L_S\{E_{PR}^p \sum\limits_{\substack{i=1,\\i\neq 4}}^{4} I_i^p\} + L_T\{E_{SR}^p(I_1^p + I_2^p)\} + L_P\{E_{SR}^p I_1^p\} - L_R\{E_{SP}^p I_3^p\}}{E_{PR}^p \sum\limits_{i=1}^{4} I_i^p}$$

(7.70)

$$l_S^3 = \frac{L_S\{E_{PQ}^p \sum\limits_{\substack{i=1, \\ i \neq 4}}^{4} I_i^p\} + L_T\{E_{PQ}^p(I_1^p + I_2^p)\} + L_P\{E_{SQ}^p I_1^p\} - L_Q\{E_{SP}^p I_2^p\}}{E_{PQ}^p \sum\limits_{i=1}^{4} I_i^p}$$

(7.71)

7.8.7.2   Optimal fault location estimation

$$l_{S,opt} = \frac{l_S^1 + l_S^2 + l_S^3}{3}$$

(7.72)

7.8.7.3   Deviations and estimation of resultant fault location estimation

$$f_1 = |l_{S,opt} - l_S^1| \quad f_2 = |l_{S,opt} - l_S^2| \quad f_3 = |l_{S,opt} - l_S^3|$$

(7.73)

From the above if f_2 is low, then $l_S^2 (= l_S^R)$ is the resultant value of fault location for fault on ST section.

7.8.7.4   Root mean error

$$f_{ST}(x) = (f_1)^2 + (f_2)^2 + (f_3)^2$$

(7.74)

$$RME_{ST} = \sqrt{\frac{f_{ST}(x)}{3}}$$

(7.75)

7.8.8   Symmetrical fault location for faults on TU section
7.8.8.1   Fault location estimations

$$l_T^1 = \frac{L_T\{E_{SR}^p(I_3^p + I_4^p)\} + L_P\{E_{RS}^p I_1^p\} + L_S\{E_{PR}^p I_4^p\} - L_R\{E_{PS}^p I_3^p\}}{E_{SR}^p \sum\limits_{i=1}^{4} I_i^p}$$

(7.76)

$$l_T^2 = \frac{L_T\{(E_{PR}^p - E_{QS}^p)(I_3^p + I_4^p)\} + L_P\{E_{QS}^p I_1^p\} + L_S\{E_{PR}^p I_4^p\} - L_Q\{E_{PR}^p I_2^p\} - L_R\{E_{QS}^p I_3^p\}}{(E_{PR}^p - E_{QS}^p) \sum\limits_{i=1}^{4} I_i^p}$$

(7.77)

$$l_T^3 = \frac{L_T\{E_{PQ}^p(I_3^p + I_4^p)\} + L_S\{E_{PQ}^p I_4^p\} + L_P\{E_{QS}^p I_1^p\} - L_Q\{E_{PS}^p I_2^p\}}{E_{PQ}^p \sum\limits_{i=1}^{4} I_i^p}$$

(7.78)

$$l_T^4 = \frac{L_T\{(E_{PS}^p - E_{QR}^p)(I_3^p + I_4^p)\} + L_R\{E_{PS}^p I_3^p\} + L_P\{E_{QR}^p I_1^p\} - L_S\{E_{QR}^p I_4^p\} - L_Q\{E_{PS}^p I_2^p\}}{(E_{PS}^p - E_{QR}^p)\sum\limits_{i=1}^{4} I_i^p}$$

(7.79)

$$l_T^5 = \frac{L_T\{E_{PQ}^p(I_3^p + I_4^p)\} + L_P\{E_{QR}^p I_1^p\} + L_R\{E_{PQ}^p I_3^p\} - L_Q\{E_{PR}^p I_2^p\}}{E_{PQ}^p \sum\limits_{i=1}^{4} I_i^p}$$

(7.80)

$$l_T^6 = \frac{L_T\{E_{RS}^p(I_3^p + I_4^p)\} - L_Q\{E_{RS}^p I_2^p\} + L_R\{E_{QS}^p I_3^p\} - L_S\{E_{QR}^p I_4^p\}}{E_{RS}^p \sum\limits_{i=1}^{4} I_i^p}$$

(7.81)

7.8.8.2 Optimal fault location estimation

$$l_{T,opt} = \frac{l_T^1 + l_T^2 + l_T^3 + l_T^4 + l_T^5 + l_T^6}{6}$$

(7.82)

7.8.8.3 Deviations and estimation of resultant fault location estimation

$$f_1 = |l_{T,opt} - l_T^1| \quad f_2 = |l_{T,opt} - l_T^2| \quad f_3 = |l_{T,opt} - l_T^3|$$

(7.83)

$$f_4 = |l_{T,opt} - l_T^4| \quad f_5 = |l_{T,opt} - l_T^5| \quad f_6 = |l_{T,opt} - l_T^6|$$

(7.84)

From the above if f_2 is low, then $l_T^2 (= l_T^R)$ is the resultant value of fault location for fault on TU section.

7.8.8.4 Root mean error

$$f_{TU}(x) = (f_1)^2 + (f_2)^2 + (f_3)^2 + (f_4)^2 + (f_5)^2 + (f_6)^2$$

(7.85)

$$RME_{TU} = \sqrt{\frac{f_{TU}(x)}{3}}$$

(7.86)

References

[1] K. Sun, Y. Hou, W. Sun, and J. Qi, *Power System Control Under Cascading Failures: Understanding, Mitigation, and Restoration*. Hoboken, NJ: Wiley, 2019.

[2] IEEE PES CAMS Task Force on Cascading Failure, Initial review of methods for cascading failure analysis in electric power transmission systems, in *Proc. IEEE PES General Meeting*, Pittsburgh, PA, 2008, pp. 1–8.

[3]   V. Rampurkar, P. Pentayya, H.A. Mangalvedekar, and F. Kazi, Cascading failure analysis for Indian power grid, *IEEE Transactions on Smart Grid*, vol. 7, no. 4, pp. 1951–1960, 2016.

[4]   D. Thukaram, H. Khincha, and B. Ravikumar, An intelligent approach using support vector machines for monitoring and identification of faults on transmission systems, in *2006 IEEE Power India Conference*, 2006, pp. 184–190.

[5]   Q. Zhao, X. Qi, M. Hua, J. Liu, and H. Tian, Review of the recent blackouts and the enlightenment, *CIRED 2020 Berlin Workshop (CIRED 2020)*, 2020, pp. 312–314.

[6]   M.M. Saha, J. Izykowski, and E. Rosolowski, *Fault Location on Power Networks*. Berlin, Germany: Springer-Verlag, 2009.

[7]   A.H. Latha and R. Bhimasingu, A novel algorithm for improving the differential protection of power transmission system, *Electric Power Systems Research*, vol. 181, pp. 106–183, 2020.

[8]   A.H. Latha, R. Bhimasingu, and S.B. Katta, A novel DC transients based differential protection scheme for transmission system, in *2018 15th IEEE India Council International Conference (INDICON)*, 2018, pp. 1–6.

[9]   B. Ravikumar, D. Thukaram, and H.P. Khincha, An approach using support vector machines for distance relay coordination in transmission system, *IEEE Transactions on Power Delivery*, vol. 24, no. 1, pp. 79–88, 2009.

[10] B. Ravikumar, D. Thukaram, and H.P. Khincha, Comparison of multiclass SVM classification methods to use in a supportive system for distance relay coordination, *IEEE Transactions on Power Delivery*, vol. 25, no. 3, pp. 1296–1305, 2010.

[11] Y.S. Meng, S.C.A. Castillo, and C.E. Arturo, Adaptive directional overcurrent relay coordination using ant colony optimisation, *IET Generation, Transmission & Distribution*, vol. 9, no. 14, pp. 2040–2049, 2015.

[12] B. Ravikumar, T. Dhadbanjan, and H.P. Khincha, Intelligent approach for fault diagnosis in power transmission systems using support vector machines, *International Journal of Emerging Electric Power Systems*, vol. 8, no. 4, pp. 272–277, 2007.

[13] A. Yadav and Y. Dash, An overview of transmission line protection by artificial neural network: fault detection, fault classification, fault location, and fault direction discrimination, *Advances in Artificial Neural Systems*, vol. 2014, December 2014, Art. no. 230382.

[14] A. Mukherjee, P.K. Kundu, and A. Das, Transmission line faults in power system and the different algorithms for identification, classification and localization: a brief review of methods, *Journal of the Institution of Engineers (India): Series B*, vol. 102, no. 4, pp.855–877, 2021.

[15] E.O. Echweitzer, A. Guzmán, H.J. Altuve, *et al.*, Real-time synchrophasor applications in power system control and protection, in *10th IET International Conference on Development in Power System Protection (DPSP)*, March 2010.

[16] M.U. Usman and M.O. Faruque, Applications of synchrophasor technologies in power systems, *Journal of Modern Power Systems and Clean Energy*, vol. 7, no. 2, pp. 211–226, 2019.

[17] A.G. Phadke and J.S. Thorp, *Synchronized Phasor Measurements and Their Applications*. New York, NY: Springer, 2008.

[18] T. Takagi, Y. Yamakoshi, M. Yamaura, R. Kondow, and T. Matsushima, Development of a new type fault locator using the one-terminal voltage and current data, *IEEE Transactions on Power Apparatus and Systems*, vol. PAS-101, no. 8, pp. 2892–2898, 1982.

[19] A. Saber, H.H. Zeineldin, T.H.M. EL-Fouly, and A. Al-Durra, A new fault location scheme for parallel transmission lines using one-terminal data, *Electric Power System Research*, vol. 135, pp. 107–548, 2022.

[20] F.V. Lopes, K.M. Dantas, K.M. Silva, and F.B. Costa, Accurate two-terminal transmission line fault location using traveling waves, *IEEE Transactions on Power Delivery*, vol. 33, no. 2, pp. 873–880, 2018.

[21] R. Syahputra, A neuro-fuzzy approach for the fault location estimation of unsynchronized two-terminal transmission lines, *International Journal of Computer Science and Information Technology*, vol. 5, no. 1, pp. 23–37, 2013.

[22] A. Salehi-Dobakhshari and A.M. Ranjbar, Robust fault location of transmission lines by synchronised and unsynchronised wide-area current measurements, *IET Generation, Transmission & Distribution*, vol. 8, no. 9, pp. 1561–1571, 2014.

[23] B. Ravikumar, D. Thukaram, and H.P. Khincha, Application of support vector machines for fault diagnosis in power transmission system, *IET Generation, Transmission & Distribution*, vol. 2, no. 1, pp. 119–130, 2008.

[24] Z.S. Chafi and H. Afrakhte, Wide area fault location on transmission systems using synchronized/unsynchronized voltage/current measurements, *Electric Power Systems Research*, vol. 197, p. 107285, 2021.

[25] D. Thukaram, H.P. Khincha, and B. Ravikumar, A new approach for fault location identification in transmission system using stability analysis and SVMs, in *Proceedings of International Conference on Power Electronics, Driver and Energy Systems*, New Delhi, India, pp. 1–6, December 2006.

[26] S. Barman and B.K.S. Roy, Detection and location of faults in large transmission networks using minimum number of phasor measurement units, *IET Generation, Transmission & Distribution*, vol. 12, no. 8, pp. 1941–1950, 2018.

[27] M.A. Elsadd and A.Y. Abdelaziz, Unsynchronized fault-location technique for two- and three-terminal transmission lines, *Electric Power Systems Research*, vol. 158, pp. 228–239, 2018.

[28] M. Silva, D. Coury, M. Oleskovicz, and E. Segatto, Combined solution for fault location in three-terminal lines based on wavelet transforms, *IET Generation, Transmission & Distribution*, vol. 4, no. 1, pp. 94–103, 2010.

[29] A. Ahmadimanesh and S.M. Shahrtash, Time-time-transform-based fault location algorithm for three-terminal transmission lines, *IET Generation, Transmission & Distribution*, vol. 7, no. 5, pp. 464–473, 2013.

[30] V.K. Gaur, B.R. Bhalja, and A. Saber, New ground fault location method for three-terminal transmission line using unsynchronized current measurements, *International Journal of Electrical Power & Energy Systems*, vol. 135, p. 107513, 2022.

[31]   T. Lin, Z. Xu, F.B. Ouedraogo, and Y. Lee, A new fault location technique for three-terminal transmission grids using unsynchronized sampling, *International Journal of Electrical Power & Energy Systems*, vol. 123, no. 12, pp. 123–135, 2020.

[32]   B. Mahamedi, J.G. Zhu, S. Azizi, and M. Sanaye-Pasand, Unsynchronised fault-location technique for three-terminal lines, *IET Generation, Transmission & Distribution*, vol. 9, no. 15, pp. 2099–2107, 2015.

[33]   M. Mirzaei, B. Vahidi, and S.H. Hosseinian, Accurate fault location and faulted section determination based on deep learning for a parallel compensated three-terminal transmission line, *IET Generation, Transmission & Distribution*, vol. 13, no. 13, pp. 2770–2778, 2019.

[34]   G. Chen, D. Cai, H. Long, *et al.*, A new scheme for fault location of three-terminal parallel transmission lines without transposer, *IET Generation, Transmission & Distribution*, vol. 15, pp. 2473–2487, 2021.

[35]   Y.H. Lin, C.W. Liu, and C.S. Yu, A new fault locator for three-terminal transmission lines using two-terminal synchronized voltage and current phasors, *IEEE Transactions on Power Delivery*, vol. 17, no. 2, pp. 452–459, 2002.

[36]   M. Abasi, A. Rohani, F. hatami, *et al.*, Fault location determination in three-terminal transmission lines connected to industrial microgrids without requiring fault classification data and independent of line parameters, *International Journal of Electrical Power & Energy Systems*, vol. 131, p. 107044, 2021.

[37]   M. Davoudi, J. Sadeh, and E. Kamyab, Transient-based fault location on three-terminal and tapped transmission lines not requiring line parameters, *IEEE Transactions on Power Delivery*, vol. 33, no. 1, pp. 179–188, 2018.

[38]   D. Lu, Y. Liu, and D. Lu, An iterative parameter-free fault location method on three-terminal untransposed transmission lines, in *2020 IEEE Power & Energy Society General Meeting (PESGM)*, 2020, pp. 1–5.

[39]   C. Purushotham Reddy and R. Bhimasingu, Synchronized measurements based fault location algorithm for three terminal homogeneous transmission lines, in *2019 8th International Conference on Power Systems (ICPS)*, MNIT Jaipur, India, (2019), pp. 1–6.

[40]   C. Purushotham Reddy and R. Bhimasingu, Fault location algorithm for three terminal homogeneous transmission lines using positive sequence components, in *2020 IEEE 9th Power India International Conference (PIICON)*, SONEPAT, India, pp. 1–6, 2020.

[41]   P.R. Chegireddy and R. Bhimasingu, Synchrophasor based fault location algorithm for three terminal homogeneous transmission lines, *Electric Power Systems Research*, vol. 191, pp. 106–889, 2021.

[42]   C. Purushotham Reddy and R. Bhimasingu, Positive sequence components based fault location algorithm for three terminal transmission network with non-homogeneous tapping, in *9th IEEE International Conference on Power Systems (ICPS 2021)*, IIT Kharagpur, India.

[43]   A. Ahmadimanesh and S.M. Shahrtash, Transient-based fault-location method for multiterminal lines employing S-transform, *IEEE Transactions on Power Delivery*, vol. 28, no. 3, pp. 1373–1380, 2013.

[44] B.K. Chaitanya and A. Yadav, Decision tree aided travelling wave based fault section identification and location scheme for multi-terminal transmission lines, *Measurement*, vol. 135, pp. 312–322, 2019.

[45] J. Ding, X. Wang, Y. Zheng, and L. Li, Distributed traveling-wave-based fault-location algorithm embedded in multi-terminal transmission lines, *IEEE Transactions on Power Delivery*, vol. 33, no. 6, pp. 3045–3054, Dec. 2018.

[46] Z. Moravej, M. Movahhedneya, and M. Pazoki, Gabor transform-based fault location method for multi-terminal transmission lines, *Measurement*, vol. 125, pp. 667–679, 2018.

[47] E.E. Ngu and K. Ramar, Combined impedance and traveling wave based fault location method for multi-terminal transmission lines, *Electric Power Systems Research*, vol. 33, no. 10, pp. 1767–1775, 2011.

[48] Y. Zhu and X. Fan, Fault location scheme for a multi-terminal transmission line based on current traveling waves, *International Journal of Electrical Power & Energy Systems*, vol. 53, no. 1, pp. 367–374, 2013.

[49] T. Funabashi, H. Otoguro, Y. Mizuma, L. Dube, and A. Ametani, Digital fault location for parallel double-circuit multi-terminal transmission lines, *IEEE Transactions on Power Delivery*, vol. 15, no. 2, pp. 531–537, 2000.

[50] A. Ghorbani and H. Mehrjerdi, Negative-sequence network based fault location scheme for double-circuit multi-terminal transmission lines, *IEEE Transactions on Power Delivery*, vol. 34, no. 3, pp.1109–1117, 2019.

[51] C.E.M. De Pereira and L.C. Zanetta, Fault location in multi-tapped transmission lines using unsynchronized data and superposition theorem, *IEEE Transactions on Power Delivery*, vol. 26, no. 4, pp. 2081–2089, 2011.

[52] S.M. Brahma, New fault-location method for a single multi-terminal transmission line using synchronized phasor measurements, *IEEE Transactions on Power Delivery*, vol. 21, no. 3, pp. 1148–1153, 2006.

[53] S. Hussain and A.H. Osman, Fault location scheme for multi-terminal transmission lines using unsynchronized measurements, *International Journal of Electrical Power & Energy Systems*, vol. 78, pp. 277–284, 2016.

[54] Q. Jiang, B. Wang, and X. Li, An efficient PMU-based fault-location technique for multi terminal transmission lines, *IEEE Transactions on Power Delivery*, vol. 29, no. 4, pp. 1675–1682, 2014.

[55] G. Manassero, E.C. Senger, R.M. Nakagomi, E.L. Pellini, and E.C.N. Rodrigues, Fault-location system for multiterminal transmission lines, *IEEE Transactions on Power Delivery*, vol. 25, no. 3, pp. 1418–1426, 2010.

[56] C.W. Liu, K. Lien, C. Chen, and J. Jiang, A universal fault location technique for N>=3 transmission lines, *IEEE Transactions on Power Delivery*, vol. 23, no. 3, pp. 1366–1373, 2008.

[57] L. Kai-Ping, L. Chih-Wen, J. Jiang, C. Ching-Shan, and Y. ChiShan, A novel fault location algorithm for multi-terminal lines using phasor measurement units, in *Proceedings of the 37th Annual North American Power Symposium*, pp. 576–581, 2005.

[58] Y. Lee, T. Lin, and C. Liu, Multi-terminal nonhomogeneous transmission line fault location utilizing synchronized data, *IEEE Transactions on Power Delivery*, vol. 34, no. 3, pp. 1030–1038, 2019.

[59] M. Abe, N. Otsuzuki, T. Emura, and M. Takeuchi, Development of a new fault location system for multi-terminal single transmission lines, *IEEE Transactions on Power Delivery*, vol. 10, no. 1, pp. 159–168, 1995.

[60] Y. Cai, A.D. Rajapakse, N.M. Haleem, and N. Raju, A threshold free synchrophasor measurement based multi-terminal fault location algorithm, *International Journal of Electrical Power & Energy Systems*, vol. 96, pp. 174–184, 2018.

[61] F. Kiessling, P. Nefzger, J.F. Nolasco, and U. Kaintzyk, *Overhead Power Lines: Planning, Design, Construction*. Berlin, Germany: Springer-Verlag, 2003.

[62] Manitoba HVDC Research Center, PSCAD/EMTDC Power System Simulation Software User's Manual, Version 4.6, https://www.pscad.com/

[63] IEEE Standard for Synchrophasor Measurements for Power Systems, IEEE Standard C37.118.1-2011, December 2011.

[64] IEEE Guide for Determining Fault Location on AC Transmission and Distribution Lines, IEEE Standard C37.114-2004, 2005.

[65] V.K. Gaur, B.R. Bhalja, and M. Kezunovic, Novel fault distance estimation method for three-terminal transmission line, *IEEE Transactions on Power Delivery*, vol. 36, no. 1, pp. 406–417, 2021.

[66] M. Pignati, L. Zanni, S. Sarri, R. Cherkaoui, J.-Y. Le Boudec, and M. Paolone, A pre-estimation filtering process of bad data for linear power systems state estimators using PMUs, in *Proceedings of Power Systems Computation Conference*, August 2014, pp. 1–8.

Chapter 8

System integrity protection schemes for future power systems

Subhadeep Paladhi[1], Kevin Kawal[2], Qiteng Hong[2] and Campbell Booth[2]

An interconnected power system must be capable of delivering electricity eco-nomically and reliably to the customers. Maintenance of all power system para-meters (voltage, frequency, real and reactive power flows, etc.) within specified limits is imperative for the effective and reliable operation of an interconnected system. A disturbance at any location, if not tackled quickly and appropriately, can aggravate a challenging situation further and may initiate cascading events, which could eventually lead to major wide-scale disruption to the power supply, as has happened several times in the past [1,2]. The rapid growth of converter-interfaced renewable generation to meet ambitious decarbonisation targets has significantly increased the complexity and uncertainty associated with future power systems, and there are growing risks and concerns relating to blackouts and system-wide disturbances, which present compelling needs to re-evaluate solutions and arrangements in place for maintaining power system integrity. Wide-area syn-chrophasor data provides a promising opportunity to system operators to identify vulnerable regions in a power system by identifying and predicting system risks/operational issues (e.g. instability), analysing system security and safety margins (e.g. overload limit), and enable effective actions to reduce the risk, and ideally avoid, a wide-spread event, thereby maintaining overall system integrity. Typically, protection relay settings may need to be adaptive to the system condi-tions to provide maximum security without compromising dependability. Corrective actions (such as load shedding (LS), generation rejection, reactive power control, etc.) may be performed to bring back the system to a desired and stable state, which could also be followed by isolating the unsynchronised areas. Individual and/or hierarchical operation of such detection and correction methods are known collectively as '*System Integrity Protection Schemes* (SIPS)'.

SIPS are also sometimes referred to differently, with terms such as special protection schemes (SPS) or remedial action schemes (RAS) [3]. Such schemes are

[1]Department of Electrical Engineering, Indian Institute of Technology Indore, India
[2]Department of Electronic and Electrical Engineering, University of Strathclyde, UK

installed to protect the integrity of a power system or a strategic portion of the system and are different from conventional protection schemes, which are dedicated to protecting specific power system elements (e.g. power lines, transformers, generators, busbars, etc.) typically against electrical faults. SIPS provide a effective countermeasure to impede and/or diminish cascading outages resulting from extreme contingencies. For this purpose, the schemes typically obtain both local and remote synchrophasor data from multiple locations using communication systems and links. Network complexity and increasing power demand, coupled with limited power transfer corridors, and minimum redundancy are the major challenges for SIPS operation [4]. Furthermore, significant changes in system dynamics and fault levels due to renewable energy integration introduce new challenges to the power system monitoring, control, and protection methods that are often incorporated within, and/or operate in parallel with, SIPS.

This chapter reviews the fundamental concepts and operating principles of SIPS with different monitoring, protection, and control methods. Classification, architecture, and design considerations in SIPS are discussed in detail in Section 8.1. Different monitoring, protection and control methods applied widely in SIPS are discussed in Section 8.2 with a hierarchical overview for SIPS implementation. Section 8.3 provides an overview of the practical implementation and operational experience of different SIPS applications in different countries. Challenges and advances in various monitoring, protection, and control methods as part of synchrophasor-based SIPS for future power systems with high penetration of renewable energy sources are discussed in Section 8.4. In Section 8.5, a number of reliability and testing requirements for SIPS are presented, along with views on current trends relating to the development of synchrophasor-based SIPS to support secure and reliable future power system operation.

8.1   Overview of SIPS

SIPS are mainly applied to provide responsive and remedial actions (such as LS, generation rejection, out-of-step (OOS) tripping, and so forth) to maintain a stable and normal state of the power system, even if the system is degraded in some way following the responsive action(s) (e.g. load is shed, generation is tripped, etc.). The concept of SIPS was first introduced by the North American Electric Reliability Corporation (NERC) [5]. NERC defined SIPS as a "Special Protection System" (SPS) which was designed to take automatic corrective actions following the detection of any abnormal condition in the power system to maintain stability and security. These actions include a change in generation/demand or system reconfiguration, but in NERC's original definition, SIPS did not include undervoltage (UV)/under-frequency (UF) LS and OOS protection. Later, the NERC standards drafting team introduced a new definition of SIPS to include the term 'Remedial Action Schemes (RAS)' replacing SPS [6]. With the predetermination of any system disorder, RAS is designed to take automatic corrective measures, which may include generation adjustment, load tripping, system reconfiguration, and other similar actions to limit the impact of extreme events by meeting the objectives under the NERC reliability standards; and maintaining bulk power system stability criteria. IEEE further revised the SIPS architecture to include different system integrity schemes (e.g. UV/UF LS, OOS protection, etc.) with SPS and RAS under the umbrella term of SIPS [3].

8.1.1 SIPS classification

SIPS are classified differently based on the employed decision-making process, the application area/architecture, the types of corrective action taken, and other aspects associated with the application. An overview of different types of SIPS is shown in Figure 8.1.

Based on the decision layers involved, the SIPS may be of two different architectures; flat or hierarchical [3]. SIPS with flat architectures involve a single layer of decisions and actions, which may require communication links to obtain remote information and to perform corrective actions. An example of a SIPS of this type is an underfrequency LS scheme, where the frequency is determined at a measurement location (normally a distribution substation) and the circuit breakers are tripped at pre-selected locations to shed load in the event of the measured frequency crossing a threshold – progressively more load is shed if the frequency continues to decline and there are normally multiple stages/thresholds of frequencies/shedding actions. A SIPS with a hierarchical architecture performs the corrective action in several steps. Examples of this type of SIPS include the

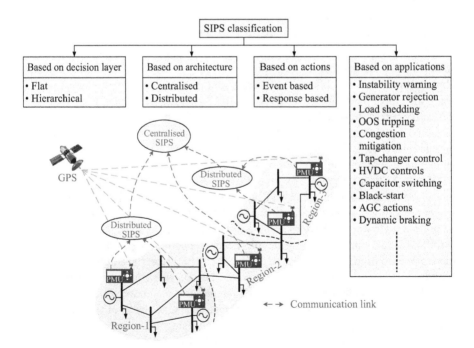

Figure 8.1 Generic architecture of SIPS

application of nomograms, state estimation and contingency analysis, which require measurements from different locations and performed at multiple control centres.

SIPS can be realised in two ways (centralised and distributed), which are largely dictated by the physical area associated with the implementation [3]. In a distributed architecture, the decision and corrective actions associated with the SIPS are implemented in different regional controllers throughout the power system as shown in Figure 8.1. With a centralised architecture, the remote information from different regions are collected at a central location through communication links and the decisions are derived centrally to perform the required action(s), which again usually requires the use of communications to implement the action(s) at remote location(s).

The corrective actions carried out by SIPS are performed following the detection of any predefined events (such as line outage contingencies and generator tripping) or abnormality detected in the measurements of different electrical parameters (such as frequency, voltage and current) [7]. Event-based SIPS are generally applied to improve transient stability or short-term voltage stability, whereas the response-based SIPS are applied for taking corrective actions such as under-frequency LS and generator rejection. SIPS typically include different applications as illustrated in Figure 8.1 [3]. Some of those applications, particularly those with significance in preserving system integrity, will be discussed later in this chapter.

8.1.2   SIPS design consideration

A generic SIPS architecture is designed with different components (such as intelligent electronic devices (IEDs), front-end processors (FEPs), controllers, workstations, operator console and archiving system [8]. SIPS operates with the data obtained from different phasor measurement units (PMUs), typically synchronised using the global positioning system (GPS). IEDs include different components (e.g. protective relays, I/O modules, remote terminal units, etc.). IEDs monitor the state of the power system, including both equipment (e.g. circuit breakers) statuses and measurements (e.g. voltage and current – and derived measurements such as frequency and phasors) and execute different control/response actions when deemed necessary. FEPs are used at the substation level for data aggregation, time alignment, and local state estimation and supply the information to the SIPS controller, where the various application algorithms are executed. A typical SIPS architecture includes a large storage drive to archive the dynamic response of the power system for a finite time duration during the pre-event to post-event period. Various parameters and configurations associated with the SIPS are set securely in the workstation. The console provides the scope to the operator for viewing the overall system human–machine interface (HMI), single-line diagrams with power flows, and the report containing the communications status and health of the devices [8]. Each substation is typically equipped with a fully integrated control device known as the substation protection controller (SPC). The SPC measures generation, aggregates data and controls digital and analogue power system assets with high-speed computation capability [8]. SPC is also programmed to perform with local data in case of communications failure.

The communication system is an essential part of any SIPS implementation. It must be designed to meet the required functional and timing requirements. SIPS signals are carried over various communication media (such as SONET, micro-wave, leased line, fibre satellite, radio, and power line carrier) [9]. The communication media for connecting a device is selected based on the reliability, latency, and bandwidth requirements specific to the SIPS. Communication links with low-bandwidth may be sufficient for transferring data such as circuit breaker status, whereas more bandwidth and lower latency may be required for transferring a large amount of analogue data and/or synchronised or time-critical data, such as for schemes using IEC 61850. Redundancy is provided in the communication system for reliable SIPS operation, even when an element of the communications system is unavailable or out-of-service.

8.2   Methods of SIPS implementation

Figure 8.2(a) shows a conceptual design for the implementation of SIPS. Such a scheme obtains wide-area synchrophasor data along with the status of planned changes in generation, load and network configuration and executes some pre-defined studies as shown in the figure. The scheme compares the present status with the predefined design and parameter requirements. If any system abnormality is identified, the scheme takes corrective measures using appropriate protection and/or control actions. A number of corrective actions performed in SIPS to mitigate several common critical power system conditions are presented in Figure 8.2(b) [8].

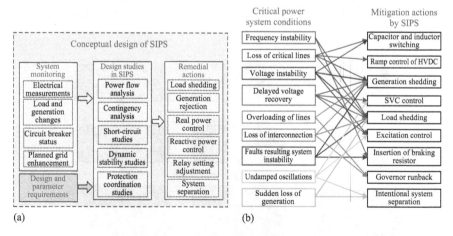

(a)   (b)

Figure 8.2   (a) Conceptual design of SIPS and (b) an overview of typical remedial actions performed by SIPS to mitigate different critical system conditions

A detailed overview of the various monitoring, protection and control techniques is discussed in this section, which are applied in SIPS to identify system abnormalities and thereafter act to maintain the integrity of the system.

8.2.1   Power system monitoring for SIPS

A power system is required to be monitored continuously to identify any abnormality threatening the system's integrity, so that proper corrective measures can be taken in advance. Two such common approaches that are commonly used in SIPS are described here.

8.2.1.1   Instability prediction

Power systems must maintain stable operating conditions, characterised by bounded system variables (voltages, frequency and rotor angles), when subject to disturbances. System stability is typically classified based on the system variable involved [10]. Rotor angle stability involves the synchronism of synchronous machines whose dynamics are governed by electro-mechanical properties. Voltage and frequency stability involves maintaining steady post-disturbance bus voltages and frequencies within tolerable ranges around nominal values. Frequency stability is influenced by the balance between generation and demand while voltage stability is dependent on adequate reactive power to support the required power flows. Varying system topologies and diverse operating conditions impede the ability of offline stability studies to fully assess the risk of system instability for all possible scenarios. Stability-based SIPS addresses such issues by providing real-time, measurement-based stability assessments which can be utilised to coordinate system wide, automatic control actions to prevent system instability. Figure 8.3 illustrates a general framework for stability-based SIPS. Stability-based SIPS leverages WAMS data available from PMU and SCADA networks to calculate appropriate metrics to assess system stability. Table 8.1 lists several common metrics used to assess different stability phenomena.

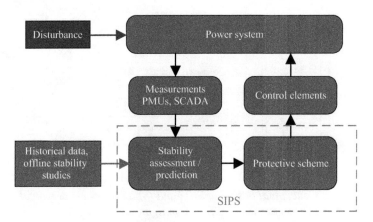

Figure 8.3   Structure of a stability-based SIPS

Table 8.1   Common power system disturbances and associated SIPS monitoring and control functions [9]

Stability phenomenon	Common stability metrics	Protection schemes
Rotor angle stability	• Generator coherency • Equal area criterion • Rotor angle separation	• Generator/LS • System separation
Voltage stability	• Power–voltage (P–V) curves • Reactive power measurements • Bus voltage measurements	• Under-voltage LS (UVLS) • Reactive power compensation • System reconfiguration • System separation
Frequency stability	• Rate of change of frequency	• Under-frequency LS schemes • Generator shedding • System separation • Spinning reserve (hydro, gas turbine, pumped storage start)

Stability criteria based on synchronous machine dynamics or power transfer limits indicate the present stability of the system as well as stability margins relative to critical thresholds. Real-time monitoring of such margins allows for SIPS to predict system stability, providing enhanced situational awareness. In [11], a metric based on generator rotor angles (position of the rotor relative to terminal voltage), measured or estimated from wide-area measurements, is proposed to assess the transient stability of an area of the network. The stability index of a power system with n generators is described by (8.1)–(8.3):

$$\delta_{avg} = \frac{1}{n} \sum_{i=1}^{n} \delta_i \tag{8.1}$$

$$\delta_{COI} = \frac{1}{M_t} \sum_{i=1}^{n} M_i \delta_i \tag{8.2}$$

$$SSI = \delta_{avg} - \delta_{COI} \tag{8.3}$$

where δ_i is the generator rotor angle of the ith generator, M_i is the inertia of the ith generator and M_t is the total system inertia. δ_{avg} represents the average system angle. δ_{COI} represents the centre of inertia (COI) angle, which is proportional to the weighted average of the individual inertia contributions of generators. During transient conditions, the synchronous machines accelerate (or decelerate, depending on the nature of the condition) and may also exhibit oscillatory behaviour, leading to changes in rotor angle. It is shown in [11] that δ_{avg} is influenced by generators with relatively lower inertia, while δ_{COI} is influenced by generators with relatively larger inertia.

The difference between these angles reflects information about the rotor angle stability of the network. During stable conditions, the stability index, SSI, shows a damped response, reflective of generators approaching a new stable operating point. In contrast, during unstable conditions, the calculated SSI increased exponentially, reflecting loss of synchronism. The magnitude and rate of change of the SSI can therefore be used to determine overall system stability. Time-series forecasting based on previous values is used to predict the trajectory of the proposed index to improve instability detection times. The predicted SSI magnitude and rate of change are then compared to thresholds which are reflective of stable conditions of the protected system. These thresholds may be determined using offline stability studies. Exceeding the thresholds is indicative of system instability. This method in [11] provides timely estimation of system stability using wide area measurements. The method requires the use of rotor angle estimation methods, knowledge of generator and system inertia and may be performed for areas of the network to reduce the computational burden.

Recent research in [12] has incorporated dynamic state estimation (DSE) and machine-learning (ML) techniques to enhance stability monitoring and prediction under real-world conditions including partial system observation and variability associated with renewable energy source (RES) output. The proposed response-based SIPS uses DSE to provide post-event rotor angle time series data which are fed to ML classification models to determine instability. Such ML models can be trained using offline simulation data and then implemented to classify the system as stable/unstable using features derived from wide area measurements. A method for voltage instability detection and monitoring utilising a maximum Lyapunov exponent (MLE)-based algorithm is presented in [13]. MLE assesses the stability of non-linear dynamical systems, such as power systems, by examining the evolution of trajectories of two starting points. Unstable systems result in the distance between the trajectories increasing exponentially. A positive MLE, therefore, indicates an unstable system. The proposed method estimates the MLE index using PMU bus voltage data to determine voltage stability in a data-driven manner. A further review of stability-based SIPS can be found in [4]. The timely assessment of system stability margins allows for targeted protection and control functions and actions that are adaptable to present system conditions. Stability prediction methods used in SIPS must therefore be computationally efficient, measurement-based and accurate for realistic system conditions.

8.2.1.2   Overload and congestion management

Power transfer capacity between different areas of a power system is typically limited due to thermal ratings of the network infrastructure: for example, transmission line ratings. System disturbances can result in large power transfers between areas of the network along transmission corridors. Overloading of such corridors can lead to further equipment damage and/or protection operations, causing further outages and increasing the threat to the stability and security of the overall network. To avoid this, system operators typically ensure that power flows along transmission corridors are limited within the capacity constraint, often using a specific margin to allow some 'headroom' for coping with contingencies. Limits may also be imposed by security standards that specify the maximum loss of infeed or loss to certain levels. Modern power networks, subject to changing demand profiles and intermittent renewable generation, will experience increased and more dynamic power flows in the future. However, increasing network capacity via network reinforcement can be economically or physically difficult or infeasible. Greater network utilisation, therefore, requires operating the network closer to its capacity limits. SIPS can function as an automated safeguard to manage overload conditions due to system disturbances (generator trips, line outages) and therefore increase the secure power transfer capacity of the network.

In [14], SIPS incorporating LS and a wide-area protection function is implemented to manage overloading and stability challenges in a network designed to operate at near full load capacity to make the best use of investment in the network infrastructure. In [15], an optimal bus splitting scheme is proposed based on a linear DC power flow model and PMU measurements. In response to overload conditions, the proposed optimisation algorithm determines the optimal bus-splitting combination to mitigate such a condition subject to the maximum power flow limits of the network elements. Such a scheme can manage overloading conditions while maintaining existing generation and load balance, thus reducing load disconnections. SIPS can therefore manage overload and congestion situations allowing for further network utilisation while meeting power system operational constraints.

8.2.2 Relay operations in SIPS

Although conventional relaying methods dedicated to elemental protection are not usually considered under the umbrella term of SIPS, any relay maloperation (which could either be spurious operation when not required to operate, or non-operation when required to operate) in a stressed condition is a major concern as mentioned in several different blackout reports [1,2,16]. Therefore, relays should be secured against any grid disturbance without compromising their dependability. Two approaches accomplishing such an objective for protective relays are described below.

8.2.2.1 Adaptive relay setting

Figure 8.4(a) illustrates a generic transmission network protected by distance relays. In a typical stepped, multi-zone distance protection scheme, backup protection is ensured in case of failure of the relay (or circuit breaker) providing primary protection of any adjacent line. With such an approach, the Zone-3 of the distance relay R_A (Z_A^{Zone-3}) should be set as mentioned in (8.4) for ensuring backup of relay R_E for a bolted 3-phase fault at bus E:

$$Z_A^{Zone-3} = Z_{AB} + 1.2Z_{BE}\left(1 + \frac{I_C + I_D + I_F}{I_A}\right) \tag{8.4}$$

where I_A, I_C, I_D, and I_F are the currents measured at bus A, C, D, and F, respectively. Z_{AB} and Z_{BE} are the positive sequence impedances of lines AB and BE. Changes in infeeds (I_A, I_C, I_D and I_F) may result in underreach/overreach issues

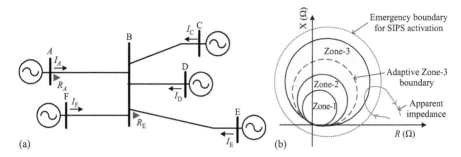

*Figure 8.4   (a) A generic transmission network protected by distance relays and
(b) adaptive Zone-3 setting-based SIPS*

when the relay is set with predefined line impedances and operates using local data.
Reduced reach settings may compromise dependability, whereas a relay may trip
unintentionally when set with a large boundary, especially when in a stressed
system condition resulting due to heavy loading, or loss of critical circuits and
generation.

To ensure the proper balance between security and dependability in a stressed
system condition, an adaptive relay setting-based SIPS is proposed in [17], which
executes two tasks simultaneously using synchrophasor data obtained from the
PMUs optimally placed in the network. The SIPS is activated when the apparent
impedance measured by a relay encroaches the emergency boundary as shown in
Figure 8.4(b), or a significant change is detected in the system configuration, such
as disconnection of lines, change in load or generation. Following the activation of
SIPS at a particular bus, a subsystem is formed starting from the relay bus.
Thevenin equivalents are estimated at the boundary buses in the subsystem, which
are further applied in the scheme to compute infeed currents for a fault condition.
Thus, the Zone-3 boundary of the relay is set adaptive to the system conditions
without compromising dependability. In a stressed condition, the apparent impedance still may encroach the Zone-3 boundary. To provide the required security for
the relay in such situations, a load prediction (LP) method is applied. The LP
technique calculates the amount of load required to be shed from a few selected
buses in the system to prevent any unintentional tripping of distance relay for such
non-fault situations. The tasks performed in this SIPS can be executed using different techniques also, such as in [18].

8.2.2.2   Power swing blocking and OOS tripping

A power system may experience severe oscillations (often in the form of power
swings as the system moves from one state to another or becomes unstable) following a large-scale disturbance (such as loss of major amounts of generation, bulk
load switching and line tripping) and a loss of balance between generation and
demand. The system may remain stable or become unstable (with generators/areas
losing synchronism) depending on the severity of oscillations. Power swings

(stable or unstable) may cause relay maloperations and possibly further loss of generation/load, leading to a cascading event. Therefore, SIPS includes special protection techniques (often within distance relays) to preserve system integrity during such oscillations [9,19].

A distance relay includes two different functions for power swing protection; power swing blocking (PSB) and out-of-step tripping (OST). The PSB function is incorporated in relays to differentiate between power swings and faults. It blocks the relay operation during a detected power swing. Monitoring the rate of change of apparent impedance is usually applied in distance relays to perform the PSB function. A relay is provided with an outer and an inner 'blinder', as shown in Figure 8.5(a) in the form of concentric circles for a distance relay with mho characteristic. A timer is placed to measure the time taken by the apparent impedance trajectory to cross both blinders. If the timer expires before the trajectory crosses the inner blinder, it declares the event as a power swing and blocks the relay operation.

To preserve system integrity during an unstable power swing, the power system may need to be divided into multiple islands by isolating a group of generators or a selected region that loses synchronism with the rest of the system during the event. Such a task is accomplished by the OST function, which is incorporated in distance relays at selected locations to differentiate between stable and unstable power swings. The locations are defined by several system studies to maintain load-generation balance in the islanded regions. There are several different techniques that may be applied to execute the OST function; a method available in [20] using synchrophasor voltage measurement is described here.

During a disturbance, the phase angle difference (δ) is computed in real time between voltages obtained from two PMUs placed on two sides of the electrical centre. Two quantities are computed (as in (8.5) and (8.6)), which are

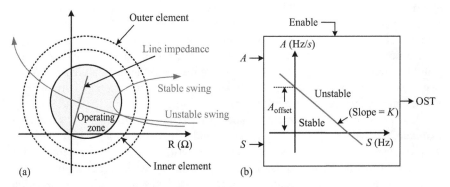

Figure 8.5   (a) *Apparent impedance trajectory encroaching distance relay operating zone during stable and unstable swing and* (b) *out-of-step trip logic using synchrophasor measurement*

the slip frequency, S (the rate of change of δ) and the acceleration A (the rate of change of S):

$$S = \frac{1}{360} \frac{\delta_i - \delta_{i-1}}{t_i - t_{i-1}} \text{Hz} \tag{8.5}$$

$$A = \frac{S_i - S_{i-1}}{t_i - t_{i-1}} \text{Hz/s} \tag{8.6}$$

where i indicates the measurement instant. The method declares the power swing to be unstable and triggers the OST function when δ is greater than a threshold, S is non-zero and increasing, and the A-S trajectory falls in the unstable region as shown in Figure 8.5(b). A combination of the PSB and OST functions described in this section is applied in [21] to divide the Uruguayan power system into two stable islands during a critical contingency caused due to the outage of one of the 500 kV lines and fault in the other remaining line. Thus, the scheme is applied to maintain the integrity of the system.

8.2.3   *Power system control approaches in SIPS*

Different control actions are performed in SIPS as a corrective measure following the detection of any abnormality in the system threatening its integrity. A list of such control actions is provided in Figure 8.2(b). The two most widely followed approaches are described here to highlight their significance in preserving system integrity.

8.2.3.1   Generation/LS

Power imbalance during contingencies can lead to overloads in the system and can negatively impact system stability. Generation and load rejection schemes are widely adopted as part of SIPS remedial actions to maintain system integrity following such imbalances during contingencies [22]. A generator rejection scheme (GRS) involves the disconnection of one or more specific generators in response to system disturbances to mitigate the risk of loss of synchronism between the selected generators and the rest of the system. This reduces the likelihood of out-of-step conditions, power swings and eventual system instability. GRSs can also reduce power flow along stressed transmission corridors thus improving system stability. GRSs are widely adopted as a SIPS remedial action for stability protection due to the fast response time in rapidly isolating generators that may participate in exacerbating potential system instability conditions.

LS refers to the curtailment of appropriate (minimised) amounts of load to maintain the balance between generation and demand during system events to prevent system collapse and sustain supply to critical loads. LS presents a 'last resort' control option in maintaining system stability when extreme disturbance conditions are being experienced. Under stressed system conditions including heavy loading, or loss of critical circuits, under-voltage LS (UVLS) can reduce power flows thus preventing voltage collapse. During frequency disturbance events, under-frequency LS (UFLS) corrects the imbalance between generation and demand and thus acts to arrest the fall in frequency, aiding in frequency recovery and preventing further system degradation. Disconnection of a fixed amount of load for predefined conditions represents the simplest form of LS. Such a static LS scheme is computationally efficient but does not consider the real-time conditions in the power system, potentially leading to over/under LS. Adaptive LS schemes determine the amount of load to be shed in accordance with system conditions and have been utilised as a remedial action in many proposed and implemented response-based SIPS [4].

Real-time stability indices derived from wide-area synchrophasor measurements can be used to determine the location and minimum amount of generation or load to be shed to maintain system integrity. A transient stability index-based SIPS is proposed in [23], which utilises the difference of voltage angles obtained from PMUs in the system to perform generation and load shedding. The amount of generation or load to be shed is determined by computing the acceleration present at the instant of the critical clearing time. The scheme is adaptive and considers communication latency, thus it can be implemented in real-time.

In [24], a GRS using predefined conditions, which are assessed based on breaker statuses from critical circuits, is implemented in a real-power network to disconnect a limited amount of generation units to avoid out-of-step conditions for the remaining units following the loss of a critical circuit. The scheme enhances transient stability in a large, interconnected power system such that the risk of power swings is reduced and the balance between generation and load is maintained. Several examples of GRS and LS are provided in [22], showing their abilities and significance in preserving system integrity.

8.2.3.2 Reactive power control

Reactive power injection can improve power transfer capabilities, manage voltage stability and prevent voltage collapse. Wider area measurement-based SIPS can monitor system voltages and determine appropriate levels of reactive power compensation during voltage events. SIPS can communicate control signals to devices, including synchronous condensers and static var compensators (SVC) to modulate the amount of reactive power delivered. Such a scheme has been implemented in [25] on a real-world network utilising both local and wide area measurements. The proposed scheme utilises existing, system-wide reactive power compensation sources to maintain voltage stability following major disturbance events in a heavily loaded part of the network. On the detection of a voltage drop, as monitored by PMUs, the data is collated at the phasor data concentrator (PDC) level, where a global control function modifies the voltage control set-points of SVCs and synchronous condensers via communication channels and a locally installed substation control unit (SCU). Local control functions are also available for SCUs in the event of communication failures. Such a distributed scheme allows for optimal and efficient use of the reactive power capacity of the network. The proposed reactive power injection-based SIPS has emerged as a more economical solution to installing new SVCs and has successfully improved the voltage stability of the grid under severe disturbance conditions.

8.3   SIPS: operational experience

PMU-enabled SIPS have been widely implemented in actual power networks throughout the world. A survey of SIPS in the Western Interconnection of the North American network in 2017 showed that there were 279 schemes in operation, with 75 incorporating wide area protection functions [26]. The survey showed that generation/load rejection schemes, in response to contingent circuit loss, are widely adopted to prevent overloading and instability of the network. An overview of the Indian power grid in [27] highlighted 52 operational SIPS as of 2020. A survey of SIPS applications used by several European TSOs as of 2016, summarised in [28], highlighted that response-based SIPS, targeting frequency and voltage instabilities, are more prominent compared to schemes targeting rotor angle instability and overload conditions. In terms of remedial actions, LS and generation rejection schemes are commonly utilised. Table 8.2 highlights several recent examples of SIPS objectives and functions implemented in power networks across several countries.

Table 8.2   Examples of SIPS implemented in real-power systems

Country	SIPS objectives	SIPS actions
New Zealand [29]	• Circuit overload protection • Transformer overload protection	• Generation shedding including windfarms • LS
Australia [30,31]	• Frequency control protection scheme • Improve interconnector transfer capabilities during peak demand	• Generation shedding • LS • Standby generation capacity provided by Battery Energy Storage system (BESS)
Belgium [32]	• Maintain system stability after loss of power corridor (leading to converter-based wind farm and HVDC links operating under weak system conditions)	• Disconnect wind power plant and HVDC link
North America [26,33]	• Circuit overload protection • Maintain system stability • Improve transfer capacities	• Generation shedding considering windfarms and coal power plants • LS • Series/shunt capacitor switching
Uruguay [34]	• Maintain voltage and frequency stability • Circuit overload protection	• LS • Reactive power compensation • Decentralised underfrequency load shedding • Generator rejection

(Continues)

Table 8.2   (*Continued*)

Country	SIPS objectives	SIPS actions
United Kingdom [35]	• Increase transfer capacities between Scotland and England • Maintain system frequency stability	• Under-frequency LS
Turkey [36]	• Maintain rotor angle stability • Prevent cascading tripping of generators • Monitor interconnector between Turkey and European network • Improve transfer capacities	• LS • Generation rejection • Reduce power transfer via HVDC link via control signals to HVDC converter controller
Taiwan [37,38]	• Maintain transient stability after loss of critical circuits • Enhance protection performance during power swing conditions	• Generation rejection • LS • Bus-tie switching • Under-frequency LS • Adaptive distance relay settings
India [27,19]	• Maintain transient stability after loss of critical circuits • Manage overloading • Enhance protection performance during power swing conditions	• Generator rejection • LS • Adaptive distance relay settings

SIPS are often employed to increase utilisation of network capacity and reduce the risk of system collapse. Recently, the proliferation of geographically dispersed RES has led to high-capacity HVDC and HVAC links being used to transfer bulk power to load centres. The uncertainty and intermittency associated with such RES can also lead to situations where there could be a need for high export of power during low-demand periods. Thus, SIPS has been implemented to ensure the safe and stable utilisation of the network under such conditions. Event-based SIPS that trigger from prespecified conditions are widely adopted, however, response-based SIPS capable of actions that are adaptive to variable system conditions in the presence of RES will likely be required in future power networks. Recently, HVDC link (and RES) converter controls [36] and fast power injection from battery energy storage systems (BESS) [30] have also been incorporated within SIPS providing new opportunities beyond generator and LS schemes. Figure 8.6 illustrates a simplified schematic of a SIPS implementation utilising reduction in power flow across an HVDC link to support a generator shedding scheme. The system consists of an area (Area 1), with high amounts of generation including bulk power transfer via an HVDC link, interconnected via two parallel transmission circuits to an area (Area 2)

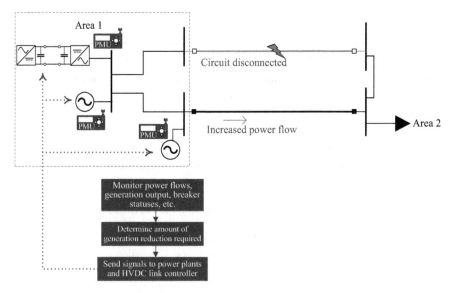

Figure 8.6   A schematic diagram of a SIPS implemented across an HVDC link

consisting of high load centres. Loss of one of the transmission circuits due to a contingency can result in increased power flow across the remaining circuit potentially leading to overloading and power swing conditions. This can result in the disconnection of large amounts of generation further threatening the overall system integrity. To mitigate this, an event-based SIPS is utilised. System information including, breaker statuses are monitored to trigger the SIPS. Specific numbers of generator units are tripped and additionally a signal from the SIPS is communicated to the back to back HVDC link to reduce the power flow to the minimum allowed level for a specified amount of time to preserve system transient stability [36].

An important practical consideration associated with SIPS implementations is concerned with ensuring dependable and secure operation. The unintended operation of SIPS or its failure to operate can have severe consequences, leading to unnecessary generation or load disconnection, propagation of system disturbances, cascading faults and eventual system instability and potential complete blackout. A SIPS maloperation in the Indian power network leading to unnecessary disconnection of generation is analysed in [27], where the SIPS was implemented to decrease generation output to prevent line overloading after a loss of one or multiple circuits emanating from the Coastal Gujarat Power Limited (CGPL) power plant. In the incident, a busbar fault occurred leading to disconnection of one circuit connecting to the CGPL power plant. Upon fault clearance a rise in voltage at the CGPL bus was interpreted as a power swing condition encroaching upon the Zone-1 characteristic of the remaining relays leading to maloperation. Post-event analysis recommended that all zones should be included as part of the monitoring conducted by the power swing blocking function. An improvement was achieved in [19] by incorporating synchrophasor measurements and deriving a stability metric using angular separation of busses to detect power swings and improve distance relay security. A review of a SIPS failure observed in the Nordic power system is analysed in [39]. The Nordic SIPS was designed to prevent overloading of transmission lines after the outage of any critical transmission corridor. When such a condition occurred, there was a delayed operation of the scheme due to communication latency, which led to a series of cascading events. The observed maloperation cases highlight the severe consequences that SIPS maloperation poses and underpin the importance of proper engineering design and reliability assessment of SIPS.

8.4   Challenges with SIPS for future power systems

The continuing and growing adoption of converter-interfaced renewable energy sources (CIRESs) and the subsequent displacement of synchronous generators (SGs) are causing a range of different challenges associated with monitoring, control and protection methods employed in SIPS, and it is important to consider these challenges within future power system scenarios.

8.4.1   Challenges in future power system monitoring

The displacement of the large rotating masses and associated system inertia (which manifests itself as relatively lower rates of changes of frequency following transient events that lead to imbalance between generation and demand) provided by large-scale SG with power electronics-interfaced CIRES (which have no natural inertia) has caused a significant reduction in the system inertia level. This leads to potentially higher levels of rate of change of frequency and generally higher levels of system volatility following such aforementioned events. Such conditions result in potentially high-impact power system events propagating through the network with greater pace and reach, threatening system stability. Converter-driven dynamics have been found to impact legacy protection functions and introduce new stability phenomena in the form of converter-driven oscillations. Sub-synchronous resonances (SSR) between CIRESs and series compensated transmission lines or weak grids are an emerging phenomenon that produces power system oscillations in the sub/super synchronous frequency range [40,41]. Several SSR events observed in actual power systems with oscillation frequencies ranging from 6 to 80 Hz have been described in [40]. Traditional PMUs may not have adequate measurement resolution to accurately capture the electromagnetic dynamics of CIRES. Figure 8.7 (a) presents a representative model used to replicate CIRES-driven oscillations under weak grid conditions observed in a real-world system [42]. Figure 8.7(b) shows a CIRES-driven oscillation simulated using the test system with the real-time digital simulation (RTDS) platform. It can be observed that PMU reporting rates can be insufficient to accurately monitor and characterise such oscillations, which can in turn be an impediment to the provision of effective mitigation actions. This example illustrates that the performance of synchrophasor-based SIPS (or

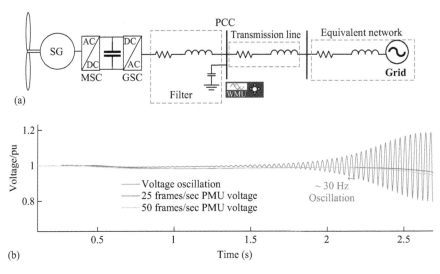

Figure 8.7   (a) A system integrating a CIRES to grid and (b) oscillation observed in a CIRES integrated system

functions within SIPS) that have been designed for power systems dominated by traditional SGs will face severe challenges in the presence of high penetrations of CIRESs.

Recent developments in power system monitoring have led to the creation of waveform measurement units (WMU) capable of producing high-resolution, time synchronised waveform (sync-wave) measurements. The enhanced monitoring resolution offered by such devices can fully capture and characterise CIRES-driven dynamics. WMU enabled SIPS, therefore, presents a promising solution to address the technical challenges introduced by CIRES. An application of WMU to real-time monitoring of CIRES-driven sub-synchronous oscillations is presented in [43]. The scheme obtains voltage and current waveforms from both ends of a series-compensated line and determines SSR current magnitude, frequency and damping within one SSR cycle. Components related to fundamental and SSR oscillation frequency are observed to be present in the waveforms that may have been inaccurately measured or filtered by traditional PMUs. Fast tracking of the oscillation damping, therefore, allows for alarming and triggering of immediate actions for oscillations with negative damping (i.e., growing oscillations). This highlights the potential for synchronised waveform measurements to enhance SIPS functionality in the presence of CIRES dynamics and provides a promising avenue for further research.

The continued integration of large-scale CIRES including offshore wind farms and solar plants have led to the investigation and installation of multi-terminal DC (MTDC) grids. The monitoring, protection, and control of MTDC grids and their interconnection with AC networks poses several emerging challenges. Different properties of conventional AC network elements and power electronic converters (high-frequency dynamics and lack of inertia contribution) on the DC side can lead to new dynamics that impact the network stability [44]. Stability analysis of MTDC grids depends on DC bus voltage magnitudes and will require different approaches compared to traditional AC networks. The stability of power electronic converters in MTDC grids are dependent on controller configuration and parameters, control modes and operating points and dynamics of both DC and AC networks [45]. This presents challenges to determining measurement-based stability metrics impeding stability-based SIPS. From a protection perspective, faults on the DC network can lead to DC voltage collapse resulting in loss of power flow which can severely impact the stability of the entire network.

8.4.2 Challenges in future power system protection

Significant reductions in fault level and system inertia, changes in voltage and current waveform shapes due to converter-control operations, and weather-dependent generation variability causing a noticeable change in infeed conditions are the major challenges for future power network protection in the presence of high penetrations of converter-based renewable sources. Some of the challenges that are related to system integrity are highlighted below.

Computation of infeed currents is necessary for proper Zone-3 setting of distance relays (as described in Section 8.2.2.1). The Thevenin equivalent of a converter-interfaced source is different during prefault and fault periods [46]. Thus, it is difficult to compute the infeeds from different buses during faults using the system equivalents estimated during prefault. Accordingly, the SIPS described in Section 8.2.2.1 cannot be applied to provide adaptive Zone-3 setting of distance relays in a converter-dominated power system.

Diversified control algorithms applied in converter-interfaced sources modulate the sequence voltage and current components differently compared to conventional synchronous generator-based sources. This may result in maloperation of the available fault type classifiers [47]. Incorrect selection of faulted phase may trigger the auto-reclosure in healthy phases, if single pole tripping is applied in the system. This is vulnerable in stressed system conditions and may initiate cascading. Fault classification techniques proposed in [47,48] may be followed to overcome such issues in the presence of converter-based sources.

The growing penetration of converter-based sources reduces system inertia significantly as outlined earlier, which eventually increases the severity and frequency of occurrence of power swings following disturbances in the system. The presence of converter-based sources influences PSB and OST functions incorporated in distance relays in different ways [49]. Reduction in system inertia increases the swing frequency significantly, which may result in maloperation of the PSB function. Control algorithms associated with converter-based sources result in a noticeable change in swing impedance trajectory. This may trigger the OST function unintentionally, even for stable power swings. Generation variability and changes in system impedance are the two major concerns in renewable-dense power systems, which influences the location of electrical centre (EC). Therefore, the location of OST relays are required to be changed adaptive to system conditions. A number of relay setting adjustment techniques are recommended in [49] to ensure correct distance relay operation during power swing. A synchrophasor assisted technique is proposed in [50], which applies the difference between the sending end and receiving end positive sequence current angles of the line for power swing detection in a wind power-dominated transmission network.

8.4.3   Challenges in future power system control

Significant reduction in system inertia due to the growing integration of converter-based sources introduces several control challenges for future power systems. A system may experience a larger rate of change of frequency (RoCoF) following a disturbance compared to a higher inertia but otherwise similar system. Maintaining the frequency within an acceptable limit in such situations, solely based on the conventional primary response from synchronous generators, is now becoming more of a challenge [51]. Non-uniform distribution of CIRESs in the system causes variation in inertia levels and frequency behaviour across the locations in the system during disturbances [51]. Frequency control of a microgrid comprising several distributed energy resources (DERs) depends on the reserve power, as well as the active power ramping rate of DERs [52]. Frequency may drop to an unacceptable level if the load shedding scheme does not consider the DERs' active power ramping capability. This may result in unintentional tripping of DERs, which could lead to total system collapse.

Several solutions are being proposed to preserve system integrity mitigating the above mentioned issues. A novel Wide-Area Monitoring and Control (WAMC) system is introduced in [51], which addresses different frequency control challenges for future power systems with CIRESs and the need for faster frequency response. The capability of such a scheme to minimise the risk of major frequency deviation incidents is demonstrated through hardware-in-the-loop testing. A novel load shedding scheme is proposed in [53] to address the challenges associated with the inadequate consideration of DERs' active power ramping capability. The scheme avoids the risk of frequency collapse in DER dominated microgrid by dynamic calculation of the amount of load needed to be shed with the consideration of DERs' reserve power as well as their active power ramping capabilities.

8.5   SIPS reliability and testing requirements for future power systems

SIPS is an integral and important part of a power system, and its performance must satisfy the same reliability requirements as all other components and systems, particularly in terms of dependability and security. Dependability has been prioritized over security in most of the adopted SIPS designs [54]. The schemes are typically designed with variable levels of redundancy and highly dependable options in terms of measurements, data communication and controllers which are responsible for deriving and issuing responsive and remedial actions when required. Power systems are becoming more complex and uncertain with the growing penetration of converter-based sources, often at distribution voltage levels (as opposed to large centralised transmission-connected sources). Any spurious operation of SIPS in such conditions may lead to cascading events. Therefore, as mentioned earlier in this section, SIPS design must maintain a balance between dependability and security [55]. Such an approach is proposed in [54] using fault tree analysis. The analysis is simplified by applying the theory of minimal cut sets, which decomposes the SIPS into individual operational phases. The minimum reliability requirement of each component is determined in terms of mean-time-to-failure and mean-time-to-fail (from a spurious operation perspective). The technique is applied to the Dinorwig Intertrip Scheme, located in North Wales, which is operated by the transmission system operator in Britain [54].

Routine testing is required for in-service SIPS to validate its performance and functionality over the system's life. Different functional and system testing is performed to monitor and ensure successful operation of SIPS [22]. In addition, SIPS design, functionality and performance should be properly validated through numerous simulation studies and hardware-in-the-loop (HiL) testing before application in the field [8]. A generalised real-time HiL test arrangement involving interfacing to SIPS controllers is shown in Figure 8.8. The SIPS are interfaced with conventional protection schemes in such a way that the SIPS are isolated automatically while under test.

Such a SIPS performance validation is found in [53], where the performance of a novel LS scheme is tested using a highly realistic HiL setup as shown in Figure 8.9. The setup consists of two parts: a real-time simulation element running in an RTDS simulator and physical systems. The LS algorithm is executed in the hardware platform included in the physical system. A modified microgrid model (available in [56]) is used for this study. The grid is emulated using a voltage

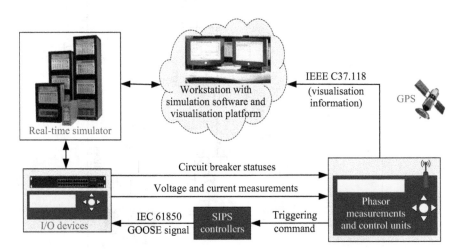

Figure 8.8   Real-time hardware-in-the loop test platform for SIPS

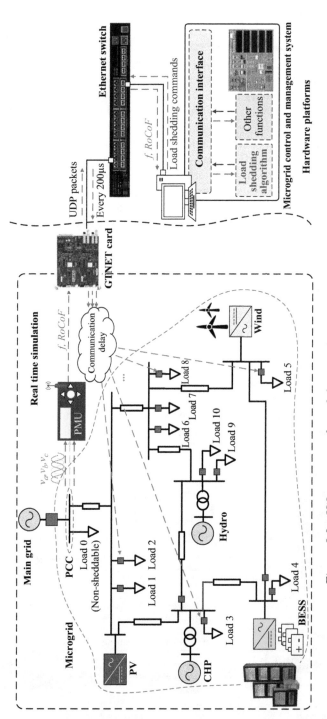

Figure 8.9 HiL test setup for validating a SIPS consisting of LS algorithm [53]

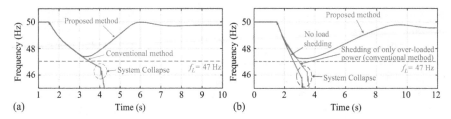

*Figure 8.10   Comparison of the performance between the proposed and
conventional LS schemes for (a) DERs' with sufficient reserve while
the ramp up rate being too slow in containing frequency deviation
and (b) both reserve power and DERs active power ramping rate are
not sufficient [53]*

source. A combined heat and power (CHP) unit, a mini-hydro plant, a battery
energy storage system (BESS), PV and wind generation units are used as DERs in
the microgrid. Total 11 loads are connected in the microgrid, in which 10 are
sheddable loads and only one is non-sheddable. Detailed descriptions of on modelling of individual component and their properties are mentioned in [53]. Under-
frequency protection is applied to each DER, where 47 Hz has been set as a
threshold with a delay of 0.5 s.

The load shedding algorithm runs individually as a part of microgrid management system. A PMU model available in RTDS obtains voltage signals from the
point of common coupling (PCC) and measures frequency and RoCoF at 50
frames/s. These are sent to the controller platform using the UDP protocol at a rate
of every 200 μs, where the LS scheme is executed. Following a requirement, the LS
commands are sent from the controller back to the RTDS to trip the required load(s)
using the UDP protocol. Different scenarios (like cases with and without sufficient
reserve power) are tested to evaluate the effectiveness of the LS scheme.

Results obtained for two case studies are presented in Figure 8.10. The first
study demonstrates the performance of the LS schemes when the DERs are with
sufficient reserve, but have slow ramp up rate. The second case demonstrates the
performance of such SIPS algorithms when both reserve power and DERs active
power ramping rate are not sufficient. It is observed from Figure 8.10 that a power
imbalance event may result in unacceptable frequency deviation in both cases and
the microgrid fails to sustain even with conventional LS schemes. On the
other hand, the step-wise adaptive LS scheme proposed in [53] helps to avoid the
violation of under-frequency threshold.

8.6   Conclusions

SIPS, the individual/hierarchical combination of monitoring, protection and control
functions, is installed in many modern power systems to preserve integrity following major contingency events. The fundamental concepts underlying SIPS have

been reviewed in this chapter, including system classifications, architectures and design considerations. Synchrophasor-based power system monitoring, protection and control schemes, which are applied widely, have been discussed in detail with a pictorial view of their hierarchical operation as part of SIPS. A list of synchrophasor-based SIPS implemented in the real world has also been provided in this chapter, highlighting the nature of applications and a number of the challenges experienced with respect to their operation. Uncertainty associated with power generation (particularly highly distributed renewables), reductions in system inertia, varying fault levels, and changes to fault behaviours and 'signatures' have all been identified as major challenges to reliable operation of future power systems. The influence of such factors on SIPS operation has been discussed individually for monitoring, protection and control functions, along with a focus on some recent advances that are intended to address a number of issues associated with present and future schemes. Finally, some coverage of how SIPS must satisfy reliability metrics, and an overview of testing requirements for ensuring dependable and secure operation of future power systems with net-zero carbon emissions, has been included. SIPS will undoubtedly play an ever-more critical role in the future as the power system continues its transition and evolution.

References

[1] Y. Besanger, M. Eremia and N. Voropai, "Major grid blackouts: analysis, classification, and prevention," in *Wiley Online Library*, New York, NY: Wiley 2013, pp. 789–863.
[2] S. Horowitz and A. Phadke, "Blackouts and relaying considerations – relaying philosophies and the future of relay systems," *IEEE Power and Energy Magazine*, vol. 4, no. 5, pp. 60–67, 2006.
[3] V. Madani, D. Novosel, S. Horowitz, *et al.*, "IEEE PSRC report on global industry experiences with system integrity protection schemes (SIPS)," *IEEE Transactions on Power Delivery*, vol. 25, no. 4, pp. 2143–2155, 2010.
[4] N. K. Rajalwal and D. Ghosh, "Recent trends in integrity protection of power system: a literature review," *International Transactions on Electrical Energy Systems*, vol. 30, no. 10, p. e12523, 2020.
[5] NERC, "Special Protection System (SPS) and Remedial Action Schemes (RAS): Assessment of Definition, Regional Practices, and Application of," Atlanta, GA, 2013.
[6] NERC, "'Remedial Action Scheme' Definition Development," Atlanta, GA, 2014.
[7] V. Madami, M. Adamiak, and M. Thakur, "Design and implementation of wide area special protection schemes," in *57th Annual Conference for Protective Relay Engineers, 2004*, 2004.
[8] K. G. Ravikumar and A. K. Srivastava, "Designing centralised and distributed system integrity protection schemes for enhanced electric grid

resiliency," *IET Generation, Transmission & Distribution*, vol. 13, no. 8, pp. 1194–1203, 2019.

[9]   "IEEE Guide for Engineering, Implementation, and Management of System Integrity Protection Schemes," *IEEE Std C37.250-2020*, pp. 1–71, 2020.

[10] P. Kundur, J. Paserba, V. Ajjarapu, *et al.*, "Definition and classification of power system stability IEEE/CIGRE joint task force on stability terms and definitions," *IEEE Transactions on Power Systems*, vol. 19, no. 3, pp. 1387–1401, 2004.

[11] P. Gawande and S. Dambhare, "New predictive analytic-aided response-based system integrity protection scheme," *IET Generation, Transmission & Distribution*, vol. 13, no. 8, pp. 1204–1211, 2019.

[12] J. A. López and C.-N. Lu, "Adaptable system integrity protection scheme considering renewable energy sources output variations," *IEEE Transactions on Power Systems*, vol. 35, no. 5, pp. 3459–3469, 2020.

[13] V. Terzija, "Data-driven and PMU-based solutions for short-term voltage instability detection and monitoring," in *2022 International Conference on Smart Grid Synchronized Measurements and Analytics*, Split, Croatia, 2022.

[14] C. De Arriba, A. Lopez, J. Cardenas, T. Woodford, and A. Bone, "Implementation of a system integrity protection scheme in the channel islands," *CIRED-Open Access Proceedings Journal*, vol. 2017, no. 1, pp. 1202–1205, 2017.

[15] Z. Zbunjak and I. Kuzle, "System integrity protection scheme (SIPS) development and an optimal bus-splitting scheme supported by phasor measurement units (PMUs)," *Energies*, vol. 12, no. 17, p. 3404, 2019.

[16] S. Paladhi and A. K. Pradhan, "Resilient protection scheme preserving system integrity during stressed condition," *IET Generation, Transmission & Distribution*, vol. 13, no. 14, pp. 3188–3194, 2019.

[17] P. Kundu and A. K. Pradhan, "Enhanced protection security using the system integrity protection scheme (SIPS)," *IEEE Transactions on Power Delivery*, vol. 31, no. 1, pp. 228–235, 2016.

[18] B. Sahoo and S. R. Samantaray, "System integrity protection scheme for enhancing backup protection of transmission lines," *IEEE Systems Journal*, vol. 15, no. 3, pp. 4578–4588, 2021.

[19] D. S. Kumar and J. Savier, "Synchrophasor-based system integrity protection scheme for an ultra-mega-power project in India," *IET Generation, Transmission & Distribution*, vol. 13, no. 8, pp. 1220–1228, 2019.

[20] N. Fischer, G. Benmouyal, D. Hou, D. Tziouvaras, J. Byrne-Finley, and B. Smyth, "Tutorial on power swing blocking and out-of-step tripping," in *39th Annual Western Protective Relay Conference*, Spokane, Washington, 2012.

[21] R. Franco, C. Sena, G. N. Taranto, and A. Giusto, "Using synchrophasors for controlled islanding – a prospective application for the Uruguayan power system," *IEEE Transactions on Power Systems*, vol. 28, no. 2, pp. 2016–2024, 2012.

[22] "Design and Testing of Selected System Integrity Protection Schemes (SIPS)," IEEE PSRC Working Group C15, 2012.

[23] S. Das and B. K. Panigrahi, "Prediction and control of transient stability using system integrity protection schemes," *IET Generation, Transmission & Distribution*, vol. 13, no. 8, pp. 1247–1254, 2019.

[24] S.-H. Lee, W.-C. Sung, J.-H. Liu, *et al.*, "Applications of system integrity protection scheme and multi-phase reclosing of transmission lines for stability enhancement in Taiwan power system," *IEEE Transactions on Industry Applications*, vol. 57, no. 5, pp. 4548–4557, 2021.

[25] E. Ghahremani, A. Heniche-Oussedik, M. Perron, M. Racine, S. Landry, and H. Akremi, "A detailed presentation of an innovative local and wide-area special protection scheme to avoid voltage collapse: from proof of concept to grid implementation," *IEEE Transactions on Smart Grid*, vol. 10, no. 5, pp. 5196–5211, 2019.

[26] "WIRAB Webinar: Introduction to Remedial Action Schemes (RAS) in the West, Western Interstate Energy Board," 20 September 2017. [Online]. Available: https://www.westernenergyboard.org/wp-content/uploads/2017/10/2017-09-29-WIRAB-RAS-Presentation-FINAL.pdf.

[27] M. S. Ballal and A. R. Kulkarni, "Improvements in existing system integrity protection schemes under stressed conditions by synchrophasor technology—case studies," *IEEE Access*, vol. 9, pp. 20788–20807, 2021.

[28] S. De Boeck, K. Das, V. Trovato, *et al.*, "Review of defence plans in Europe: current status, strengths and opportunities," *CIGRE Science & Engineering*, vol. 5, pp. 6–16, 2016.

[29] Transpower, "Special Protection Systems," [Online]. Available: https://www.transpower.co.nz/system-operator/information-industry/operational-information-system/special-protection-schemes. [Accessed 11 September 2022].

[30] "Validation Business Case for a Victorian SIPS Service," PricewaterhouseCoopers (PwC), 2020.

[31] "Technical Parameters of the Tasmanian Electricity Supply System, Information Paper," Electricity Industry Panel Secretariat, 2011.

[32] R. Hanuise and C. Moors, "Ensuring the stability of the Belgian grid with a special protection system," in *47th Annual Western Protective Relay Conference*, 2020.

[33] R. Shah, R. Hines, and P. Palen, "Upgrading PacifiCorp's Jim Bridger RAS to include wind generation," in *48th Annual Western Protective Relay Conference*, 2021.

[34] J. Malcón and N. Yedrzejewski, "Implementing a country-wide modular remedial action scheme in Uruguay," in *42nd Annual Western Protective Relay Conference*, 2015.

[35] C. McTaggart, J. Cardenas, A. Lopez, and A. Bone, "Improvements in power system integrity protection schemes," in *10th IET International Conference on Developments in Power System Protection (DPSP 2010). Managing the Change*, 2010.

[36] F. Iliceto, Ö. Akansel, M. Akdeniz, S. Erikci, Y. Z. Korkmaz, and K. Ravikumar, "System integrity protection schemes in the 400 kV

transmission network of Turkey," in *7th International Conference on Power System Protection and Automation*, 2018.

[37] C.-J. Liao, Y.-F. Hsu, Y.-F. Wang, S.-H. Lee, Y.-J. Lin, and C.-C. Chu, "Experiences on remediation of special protection system for Kinmen power system in Taiwan," *IEEE Transactions on Industry Applications*, vol. 56, no. 3, pp. 2418–2426, 2020.

[38] S.-H. Lee, W.-C. Sung, J.-H. Liu, *et al.*, "Applications of system integrity protection scheme and multi-phase reclosing of transmission lines for stability enhancement in Taiwan Power System," *IEEE Transactions on Industry Applications*, vol. 57, no. 5, pp. 4548–4557, 2021.

[39] N. Liu, Reliability Assessment of a System Integrity Protection Scheme for Transmission Networks, The University of Manchester (United Kingdom), 2018.

[40] Y. Cheng, L. Fan, J. Rose, *et al.*, "Real-world subsynchronous oscillation events in power grids with high penetrations of inverter-based resources," *IEEE Transactions on Power Systems*, vol. 38, no. 1, pp. 316–330, 2023, doi: 10.1109/TPWRS.2022.3161418.

[41] X. Xie, W. Liu, J. Shair, C. Dai and X. Liu, "Subsynchronous control interaction: real-world events and practical impedance reshaping controls," in *The 10th Renewable Power Generation Conference (RPG 2021)*, 2021.

[42] H. Liu, X. Xie, J. He, *et al.*, "Subsynchronous interaction between direct-drive PMSG based wind farms and weak AC networks," *IEEE Transactions on Power Systems*, vol. 32, no. 6, pp. 4708–4720, 2017.

[43] B. Gao, R. Torquato, W. Xu, and W. Freitas, "Waveform-based method for fast and accurate identification of subsynchronous resonance events," *IEEE Transactions on Power Systems*, vol. 34, no. 5, pp. 3626–3636, 2019.

[44] J. A. Ansari, C. Liu, and S. A. Khan, "MMC based MTDC grids: a detailed review on issues and challenges for operation, control and protection schemes," *IEEE Access*, vol. 8, pp. 168154–168165, 2020.

[45] P. Rodriguez and K. Rouzbehi, "Multi-terminal DC grids: challenges and prospects," *Journal of Modern Power Systems and Clean Energy*, vol. 5, no. 4, pp. 515–523, 2017.

[46] S. Paladhi and A. K. Pradhan, "Adaptive distance protection for lines connecting converter-interfaced renewable plants," *IEEE Journal of Emerging and Selected Topics in Power Electronics*, vol. 9, no. 6, pp. 7088–7098, 2021.

[47] S. Paladhi and A. K. Pradhan, "Adaptive fault type classification for transmission network connecting converter-interfaced renewable Plants," *IEEE Systems Journal*, vol. 15, no. 3, pp. 4025–4036, 2021.

[48] A. Hooshyar, E. F. El-Saadany, and M. Sanaye-Pasand, "Fault type classification in microgrids including photovoltaic DGs," *IEEE Transactions on Smart Grid*, vol. 7, no. 5, pp. 2218–2229, 2016.

[49] A. Haddadi, I. Kocar, U. Karaagac, H. Gras, and E. Farantatos, "Impact of wind generation on power swing protection," *IEEE Transactions on Power Delivery*, vol. 34, no. 3, pp. 1118–1128, 2019.

[50] J. T. Rao, B. R. Bhalja, M. V. Andreev, and O. P. Malik, "Synchrophasor assisted power swing detection scheme for wind integrated transmission

network," *IEEE Transactions on Power Delivery*, vol. 37, no. 3, pp. 1952–1962, 2022.

[51]   Q. Hong, M. Karimi, M. Sun, *et al.*, "Design and validation of a wide area monitoring and control system for fast frequency response," *IEEE Transactions on Smart Grid*, vol. 11, no. 4, pp. 3394–3404, 2020.

[52]   Q. Hong, M. Nedd, S. Norris, *et al.*, "Fast frequency response for effective frequency control in power systems with low inertia," *The Journal of Engineering*, no. 16, pp. 1696–1702, 2019.

[53]   Q. Hong, L. Ji, S. M. Blair, *et al.*, "A new load shedding scheme with consideration of distributed energy resources' active power ramping capability," *IEEE Transactions on Power Systems*, vol. 37, no. 1, pp. 81–93, 2022.

[54]   M. Panteli, P. A. Crossley, and J. Fitch, "Design of dependable and secure system integrity protection schemes," *International Journal of Electrical Power & Energy Systems*, vol. 68, pp. 15–25, 2015.

[55]   N. Liu and P. Crossley, "Assessing the risk of implementing system integrity protection schemes in a power system with significant wind integration," *IEEE Transactions on Power Delivery*, vol. 33, no. 2, pp. 810–820, 2018.

[56]   "Microgrid modelling and simulation applications manual," RTDS Technologies, Winnipeg, Canada Inc., 2016.

Chapter 9

Application of synchrophasor technology for microgrid protection

Abhisek Mishra[1] and Premalata Jena[1]

In this chapter, the synchrophasor technology for microgrid protection is discussed. In the present modern grid, the complex microgrid structure demands more attention towards developing fast, efficient, and reliable protection schemes to deal with various faults, and islanding events. In this regard, synchrophasor technology plays a major role due to its ability to measure the time-stamped microgrid data. Therefore, a survey on the synchrophasor technology is carried out and its application for microgrid protection is discussed in this chapter. First, the basic operation of synchrophasor technology is provided, and then, an overview of microgrid protection is discussed. Subsequently, the recent developments of the phasor measurement unit (PMU) application towards microgrid protection are provided, and at last, the possible future scope to overcome the issues associated with this technology is discussed.

9.1 Introduction

With the present modernization and the installation of smart devices in the microgrid environment, the complexity of the power distribution networks keeps on increasing rapidly. Moreover, the presence of distributed energy resources and other active loads inside the microgrid makes the distribution network more complex with faster dynamics. In this regard, the protection of microgrids against various disturbances has been an utmost priority, which requires constant monitoring of the system parameters such as bus voltages, line currents, operating frequencies, phase angle variations, power factors, active power, and reactive power. Due to the penetration of renewable sources and active loads, the distribution system is changed from a slower changing radial network to a multi-source network with faster dynamics. In this aspect, many researchers have started the work towards developing synchrophasor-based technology for fast, accurate, and reliable operation of distribution networks in the presence of distributed generation. The

[1]Department of Electrical Engineering, Indian Institute of Technology Roorkee, India

first phasor measurement system deployed in a distribution system, known as FNET/GridEye was developed in 2003. The frequency disturbance recorders (FDRs) used in it to record system data such as voltage, phase angle, and frequency, has high accuracy at a low cost. Subsequently, this develops attention of the researchers to develop wide-area monitoring system (WAMS) for the distribution level. Micro-PMU also helps to improve the observability and control of the distribution system.

In the last decade, the synchrophasor device or the PMU has been proven to be an efficient tool, as far as monitoring and control of the system parameters are concerned. For the distribution networks, the high precision synchrophasor device, known as the micro-PMU (μ-PMU) was first developed by the Power Standard Lab with the collaboration of the University of California and the Lawrence Berkeley International Laboratory. The μ-PMUs are specially developed for microgrid application, where the precision is as high as $\pm0.05\%$ of total vector error. The μ-PMU can read and stream the data at the rate of 100–120 readings/s for 50 or 60 Hz-based applications. They have a sampling rate of 25,600 samples/s for 50 Hz and 30,720 samples/s for 60 Hz system. Usually, the μ-PMUs have eight differential voltage inputs with the specification of range 0.333 Vrms or 10 Vpk, input impedance 33.3 kΩ, and crest factor 3.5. The 3 phase current channels streamed via IEEE C37.118-2 communication protocol. These devices have both P-class and M-class applications and are compliant with IEEE C37.118.1-2011 and C37.118.2-2011. These devices have the angle accuracy level of $\pm0.01°$ and a magnitude resolution in the order of $\pm0.0002\%$. With all of these above-mentioned high precisions of the μ-PMU, certainly, these are the best monitoring devices which can handle the faster dynamics of the microgrid environment. These devices are built on existing power quality disturbances recorders which can transmit the data in real-time. The devices can measure phase angle with a resolution and accuracy of $0.001°$ and $0.01°$ respectively. Resolution is defined as the smallest change in the measured quantity that can cause a detectable change in the output of the instrument. In μ-PMU, the phase angle resolution is specified as $0.001°$. Hence, the minimum value of phase angle variation that can be detected by μ-PMU is $0.001°$. The device can be connected at primary or secondary distribution voltage levels both for single-phase and three-phase applications. Communications are handled in μ-PMUs using ethernet or cellular modem. Microgrid protection includes protection against various types of faults and islanding conditions.

In the last two decades, many researchers have proposed different techniques to secure the microgrid in the event of these unwanted conditions. In this aspect, the μ-PMUs have played a vital role. The algorithms installed at the terminal of fault detection relays and islanding detection relays need constant monitoring of the system parameters, which can be obtained from the μ-PMUs. The parameters obtained from μ-PMUs are synchronized with each other by the time-stamping mechanism with the help of GPS clock. The synchronized data certainly helps in taking error-free and reliable decisions in a microgrid. Therefore, the importance of μ-PMUs in microgrid protection has gained a lot of attention in the research community. The applications of μ-PMUs include event detection and classification, topology and cyber-attack detection, model validation for loads and DGs, DG characterization, microgrid operation, distribution system state estimation, and μ-PMU measurement data applications for DG control.

It is observed that the power distribution systems have limited observability and diagnostic capabilities. This is for good reason that the earlier distribution network was radial in nature (i.e., unidirectional power flow) and equipped with little instrumentation beyond the substation. But at the same time, the growth of RES such as solar photovoltaics, rotating machine-based generating systems, automated demand response, distributed storage, and active load (i.e., electric vehicles) introduces significant variability, fast dynamic situations, and uncertainties and asks for adaptive technologies for reliable operation. This scenario dramatically raises the need for robust and reliable techniques to better observe, understand, and manage the grid at the distribution scale. The grid integration of renewable sources could lead to a more challenging platform for monitoring and protection of the network and which may not be achieved by considering only voltage and current magnitudes, but phase angles as well. The conventional techniques based on current information available at the control center are insufficient to handle all uncertainties and dynamics entrenched due to the integration of RES, active loads, and storage devices. Therefore, there is a strong need for new technology for a wide range of system monitoring, control, and protection with consideration of both the magnitude and the angle of voltage and current signals in the distribution management system. This situation could encourage to incorporate the data coming from the PMU into the functions of DMS (distribution management system) in a better way to meet the requirement.

9.2 Basic operation

As the focus of this chapter is to discuss the application of PMU toward microgrid protection, the basic working principle of PMU is discussed briefly in this section. In Figure 9.1, the block diagram of PMU operation is provided. At first, the voltage and current samples are obtained and fed to the anti-aliasing filters. The high-frequency harmonic components of the input signals are filtered out by the anti-aliasing filters. It should be noted that the sampling frequency of the anti-aliasing filters should be

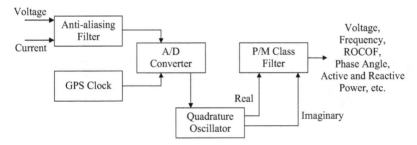

Figure 9.1   Basic block diagram of PMU block

chosen in such a way that, it should satisfy the Nyquist criteria, i.e., the sampling frequency should be more than twice the cut-off frequency. The obtained signals are then passed through the analog-to-digital converters (ADCs), which convert the analog inputs to digital form. For PMU applications, the 12-bit or 16-bit ADCs are used. These signals are time-stamped using the GPS clock signal. The output of the ADCs is then multiplied with the "*sine*" and "*cosine*" functions generated by the quadrature oscillators. The real and imaginary parts of the input signals are further passed through the P-class and M-class filters, from where the final outputs, such as voltage, frequency, active and reactive powers, rate of change of frequency (ROCOF), and phase angles, are obtained. To estimate the phasors and frequency, various phasor and frequency estimation techniques have been used in the literature.

The voltage and current signals can be represented as a sinusoid function, which can be represented as

$$f(t) = f_m.\cos(\omega t + \theta) \tag{9.1}$$

Now, (9.1) can be represented in phasor form as expressed in (9.2)

$$f(t) = \frac{f_m}{\sqrt{2}} e^{j\theta} = \cos\theta + j.\sin\theta = f_{real} + f_{img} \tag{9.2}$$

where f_m is the magnitude, $\frac{f_m}{\sqrt{2}}$ is the RMS value, θ is the phase angle of the input signal. f_{real} and f_{img} are the real and the imaginary of the complex representation.

In Figure 9.2, a layout of a phasor measurement system is provided. The PMUs are installed at various parts of the power system networks. The PMUs monitor the grid parameters (positive sequence voltage and currents of different buses and feeders [1]) and store it locally in the local storage units. These data are time-tagged by the GPS clock signal. Next, these data are transmitted to the phasor data concentrators (PDCs). The PDCs gather the data of all the system PMUs and create a database with respect to the tagged time-stamping. Any bad data present in the database are also eliminated at this stage. The network operator at the energy management system collects these data for further processing. The collected data can be monitored, controlled, and stored as per the requirement of the network operator.

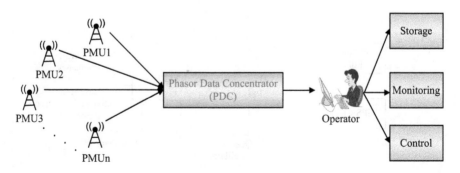

Figure 9.2   Layout of phasor measurement system

9.3   Application of PMU for microgrid protection

The synchronized measurement can be applied to various domains of power systems, such as monitoring, protection, and control of power grids [2]. The monitoring application includes voltage, frequency monitoring, and system state estimation. The PMU application in power system protection includes backup protection, out-of-step protection as well as security dependability. Similarly, the PMU can be applied for power system control through various HVDC controllers and FACTS controllers. Broadly, the authors in [3] have classified the key PMU benefits as wide-area monitoring, voltage instability monitoring, oscillation detection, generator model and parameter validation, islanding detection, event analysis, and identification of potential malfunction of devices in the grid. Out of the large applications of PMUs, this chapter focuses on the application of PMUs for microgrid protection. In this aspect, first, the introduction of microgrid protection is discussed and subsequently, the PMU layout for the monitoring and protection of microgrid is described.

9.3.1   Microgrid protection

Microgrid protection is a challenging task, which involves protection against fault as well as protection against islanding. Any major disturbances in the system may result in faults inside the microgrid. During any major islanding event, when a certain portion of the network gets disconnected from the grid and operates independently with the help of local distributed energy resources (DERs), then it is called an islanding event.

Islanding can be classified as intentional islanding and unintentional islanding. In intentional islanding, small islands are formed as the last means to avoid a system-wide blackout in the event of a disturbance in the power system network. This mode is commonly used during maintenance. Unintentional islanding is usually caused by unpredictable disturbances in the utility. If not detected in time, they may cause several problems such as safety problems for repair crews, damage to the end-user equipment as well as utility equipment, and power quality degradation. Islanding events, caused by any major faults in a microgrid, must be detected within prescribed times. A lot of work has already been done for fault location identification and removal of faults. The faulty sections must be isolated and service must be provided to the loads in the blackout area as soon as possible. Similarly, if an islanding event occurs in a microgrid, the islanded area must be detected and then, it must be restored back to the main grid. During any emergency condition, intentionally islanding is used as a last resort to keep the power supply to critical loads from the DG side.

After the emergency condition or maintenance, the islanded area should be connected back to the grid as before. The restoration in a power system network can be solved either by using build-up or build-down strategies [4]. In the case of build-down strategies, the whole network is energized as bulk and then, the restoration of load and generation is carried out step-by-step. But this method suffers from the excess reactive power generation in the unloaded high voltage system. Due to the availability of DERs inside low-voltage microgrids, it is more adequate to use build-up strategies by sectionalizing the whole microgrid network into several smaller islands and then resynchronizing them together with the grid.

A microgrid operation is shown in Figure 9.3. In this figure, the red color indicates the disconnected or non-operational elements in the system, whereas the black and green color indicate operational elements in the system. In Figure 9.3(a), the healthy microgrid before the inception of fault is shown. Suppose, a fault is incepted near the grid. After the detection of the fault, the circuit breaker "CB" is tripped out and the downstream side of the "CB" gets islanded unintentionally as shown in Figure 9.3(b). Next in Figure 9.3(c), the unintentional islanding event is detected by the islanding detection relays (IDRs) and they disconnect the DERs to supply the loads, as shown in Figure 9.3(d). Now in Figure 9.3(e), with the absence of power supply from the main grid and DERs, the system undergoes a complete system-wide blackout event. Now, the next target is to form restore back the microgrid to its healthy operating condition. In this regard, the build-up restoration strategy is used due to the availability of DERs in the microgrid environment. In the first step, smaller and stable islands are formed by the DERs of the microgrid by supplying local loads as per their supply capacities. This is shown in Figure 9.3(f). It should be noted that the islands, formed in this step, are intentional islands. As the next step of the build-up restoration strategy, the smaller islands are reconnected together to form a bigger and more stable island, as shown in Figure 9.3(g). When the fault is cleared, the reconnection of the main grid with the stable island is carried out in order to restore the healthy operation of the microgrid. This is shown in Figure 9.3(h).

9.3.2   PMU-assisted microgrid protection layout

In Figure 9.4, the layout of the PMU-assisted microgrid protection scheme is provided. Various communications in the system are indicated along with their directions. The μ-PMUs are installed at different locations of the system, which are synchronized by the GPS clock signals. This is indicated by the "communication 1" lines. The synchronization is unidirectional, i.e., from GPS to μ-PMU. The synchronized data obtained by various μ-PMUs are sent to the energy management system (EMS), which is indicated by the "communication 2" lines. The μ-PMU data are then collected for further processing. In the event of any system-wide disturbances or fault scenarios, the trip signal is generated to isolate the faulty section of the microgrid. This trip signal communication is indicated by the "communication 3" lines. Similarly, during the event of restoration process, the communication is made from the energy management system to reconfigure the microgrid structure through the "communication 3" lines.

In Figure 9.5, the advantage of using a PMU-assisted microgrid protection scheme is shown. The figure represents the frequency difference-based islanding detection scheme. For a given microgrid network, if the islanding detection rule is set as if the frequency difference between the DER side (f2) and grid side (f1) is more than 1.1 Hz, then an islanding event is detected. In the upper half of the figure, at any instant of time f1 − f2 is always less than 1.1 Hz. Therefore, no

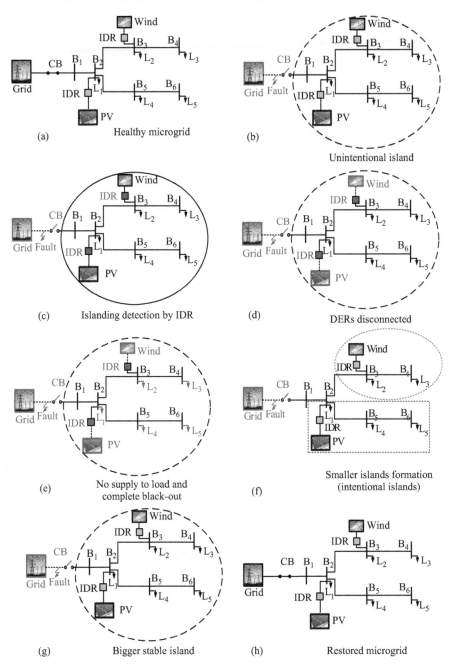

Figure 9.3 Microgrid operation

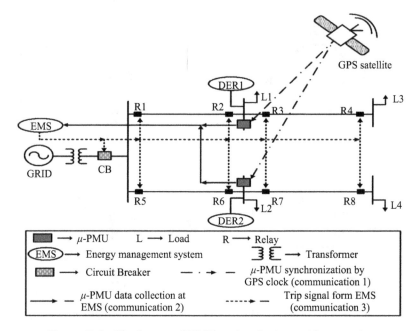

Figure 9.4   The layout of PMU-assisted microgrid protection

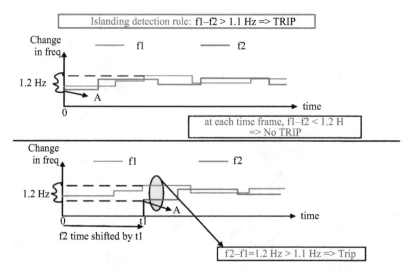

Figure 9.5   Advantage of PMU-assisted microgrid protection

islanding event is detected. The lower half of the figure indicates that any communication delay caused by the communication channel may force erroneous decision. Suppose, the frequency of the DER side (f2) is received at the energy management system by a certain time delay.

If both f1 and f2 are not time-stamped properly, the delayed recording of f2 may satisfy the islanding detection condition and mal-operate the islanding detection relay, where in reality there is no islanding event that exists inside the microgrid. Therefore, it is evident that the synchronized and time-stamped measurement can eliminate the issue of false detection and in this aspect, the application of PMUs can certainly provide reliable protection.

9.4   Literature survey

There exist many research works in the literature where the researchers have used synchrophasor technology for microgrid protection. In the last decade, the development of synchrophasor technology attracts the attention of various researchers for power system monitoring, protection, and stability aspect due to its ability for time synchronization between various system parameters which in terms eliminates the need for external synchronizing units [5,6]. Initially, the synchrophasors, alternatively known as PMUs, are used at the transmission level. Due to the availability of DERs and active loads in the distribution system, the dynamics of the system change faster, which calls for more accurate measurements for the safe operation of the grid. Moreover, signal estimation in the distribution level is a difficult task due to the error caused by negligible power angels, high levels of harmonics and noise, and smaller line lengths. μ-PMU helps to estimate signals at the distribution level due to its higher resolution of 120 readings/s [8]. In this regard, many researchers have started the work towards developing synchrophasor-based technology for fast, accurate, and reliable operation of the distribution system in the proximity of DERs [7]. The PMUs, which are used in distribution levels, are usually referred to as μ-PMU. These islanding detection methods (IDMs) can be classified into two types: central and local IDMs. In central IDMs, first, the μ-PMU data are obtained at the central controller and the islanding decision is taken at the central controller based on the algorithms installed. Once the islanding is detected, then the trip signal is sent from the central controller to the islanding detection relays. Therefore, it can be said that centralized schemes require two-way communications. In the case of local IDMs, the islanding detection is taken locally by using only local μ-PMU data. Contrary to centralized decision-making schemes, local schemes require only one-way communication since islanding decision is taken locally at the DG terminal itself.

For detection of islanding events, a principal component analysis (PCA)-based approach is adopted in [9] using synchrophasor data. In this research work, the advantage of recursive PCA with respect to normal PCA is also analyzed. In [10], synchrophasor data is used to identify islanding events using frequency and angle differences at the frequency disturbance recorder unit. But the technique takes a longer duration to detect the islanding event. In [11], the authors have proposed the application of synchrophasor data for different passive IDMs in a real-time hardware-in-the-loop (RT-HIL) environment. It is observed that the NDZ of the proposed scheme is half or two-thirds of the NDZ of local-based IDMs. In [12], the synchronized data are used to detect islanding events in the microgrid by applying the multi-resolution Teager energy operator. The dynamic modeling of microgrid and islanding control are provided in [13]. In [14], the authors have used μ-PMU data to detect islanding events inside a microgrid with the help of Pearson's coefficient. From the μ-PMU data of reference substation and DER terminals, two parameters are formulated: a cumulative sum of frequency difference and a phase angle difference. A Pearson's correlation coefficient-based islanding detection parameter is formulated to detect islanding. In [15], a peak-ratio analysis method (PRAM)-based islanding detection method is proposed. In this work, the ROCOF signal is first obtained from the PMUs installed at the DER bus terminal. Then, the ratio of ROCOF peak values on both sides of the zero axis is evaluated to take the decision. The proposed method can detect islanding events for the active power mismatch of 2.5%. In [16], a multi-function algorithm to detect islanding, faults, and power swings is proposed. However, the scheme requires μ-PMU at all buses, which is not economical. In [17], an IDM using PMU data is proposed where the phase angle difference between DER bus and reference bus is used to detect islanding event in the microgrid. The authors in [18] have proposed a scheduled intentional islanding method using synchrophasor data. In this work, the microgrid is islanded intentionally to secure the critical loads based upon a ranking approach. The possible islanding zones of the microgrid are ranked based on the overall stability measurement. In order to rank the islanded zones, the stability margins with respect to bus voltage, frequency, line flow as well as weighted loads inside the islanded zones are considered.

Synchrophasor technology has also been applied efficiently to identify and isolate faults in the microgrid. In [19], the synchrophasor data are used to locate the fault for a three-terminal hybrid transmission line. The authors in [20] have proposed a PMU measurement-based method for fault location for an untransposed and meshed transmission network. In [21], the authors have used the synchrophasor data for microgrid protection using an integrated impedance angle scheme. In [22], a supervised learning approach is used to localize fault inside a microgrid using PMU data. The technique can efficiently isolate the fault before the critical time for the safe operation of the microgrid. In [23], the PMU data is used to obtain the composite magnitude-phase plane from the impedance difference, which is used for microgrid protection against fault. In [24], the μ-PMU is utilized to locate the fault in an active distribution network. The μ-PMU location is used for dividing the microgrid into various protection zones. The variation in positive sequence line current is observed by the μ-PMU continuously, which helps in locating the fault zone. The proposed scheme is applicable for both grid-connected as well as the islanded network.

As far as the restoration problem is concerned, synchrophasor technology has been proven as an efficient tool in recent days. In [25], the synchrophasor technology is used to propose a two-stage restoration technique at the transmission level. The first stage targets to allocate loads in the system whereas the second stage maintains the overall system stability. In [26], the frequency and rate of change of frequency measurements from the synchrophasor devices are used for load restoration as well as under-frequency load shedding algorithms. In [27], a multi-agent scheme for fault location, isolation as well as service restoration scheme is proposed using the synchrophasor data. The phase angle difference between the currents at both sides of the feeder is being used to detect and locate the fault. In the next stage, the service restoration is performed by keeping the voltage, frequency as well as line constraints of the distribution system. In [28], the ROCOF measurement algorithms' impact on the grid restoration is analyzed with reference to the under-frequency load shedding. The authors in [29] have used the PMU measurements for grid restoration through the intentionally controlled islanding concepts using a mixed integer linear programming model. In [30], a Kalman filter-based dynamic state estimation is used to estimate frequency from the PMU measured data and subsequently used for frequency restoration of a PV-interfaced power system. A conservation voltage reduction strategy-based restoration scheme is proposed in [31], where the optimization problem is solved using a mixed-integer quadratic constraint programming technique. The PMU data is used to design a correction table, which decides the load-shedding plan in the system.

The optimal placement of PMUs inside a power system network is another aspect to be considered in order to have better observability, i.e., the optimal placement of PMUs ensures lesser PMU requirement for wider yet better observation of the network. Subsequently, this reduces the overall cost as well as network complexity as well. There are several techniques available for allocating PMUs optimally inside a power system network. Broadly, they are classified into Heuristic Optimization methods and Deterministic methods [35]. The heuristic approaches mainly incorporate intelligent search algorithms, whereas, in deterministic approaches, the extensive application of numerical methods is carried out. The heuristic methods such as Genetic algorithm [36,37], Particle Swarm Optimization [38], and Tabu search algorithm [39], and deterministic approaches such as integer linear programming [40] and binary search algorithm [41] are already available in the literature. In an active distribution network, number of techniques are also available for optimal placement of μ-PMUs. In [42], a usable zero-injection phase strings-based approach is incorporated in the distribution network. In hybrid methods, the global search algorithm [43] and graph theory/greedy search algorithm [44] are also incorporated in microgrid level in order to obtain the optimal location of the μ-PMUs. In [45], a tri-objective optimal PMU placement technique is proposed, which targets to minimize number of PMU channel, maximum state estimation uncertainty as well as sensitivity with respect to line tolerance.

9.5 Future research scope

In this section, the possible research directions with respect to microgrid protection using synchrophasor technology are discussed. As discussed in previous sections, the PMU records the data at the downstream side of the microgrid. For remote protection techniques, these data are transmitted to the energy management system for further data processing and decision-making. This data transmission may add time latency to the PMU measurement. Time latency is caused due to the time delays that occur in the synchrophasor data processing. Moreover, the length of the data window required for frequency estimation techniques is more than that of the phasor estimation techniques. This results in more latency in time. While the local protection schemes may not get affected by this PMU data transmission, however, the time delay of this data transmission may affect the remote protection scheme, which should be taken into consideration while designing the protection scheme.

Another concerning factor that keeps researchers busy in recent days is the "missing data points." Due to any missing data points in PMU measurements, the microgrid protection scheme may provide erroneous results and force a system-wide blackout in the microgrid. In this regard, the data recovery of the PMUs should be done in order to avoid such issues. Recently, many researchers have been working in the area of data recovery in μ-PMUs. In [32], the authors have proposed a data recovery technique in PMUs using Online Algorithm for PMU data processing (OLAP) technique. In [33], the authors have used an improved cubic interpolation method for PMU data recovery. The authors in [34] have proposed a PMU data recovery technique using low-rank matrix completion methods. During the development of any protection scheme, the researchers should investigate the impact of such data recovery techniques on the proposed protection scheme.

"Bad data" is another issue associated with PMU data transmission. Any malfunction in the communication infrastructure or presence of noise during communication may result in "Bad data." While developing any microgrid protection scheme, more emphasis should be given to the presence of bad data. The developed algorithm should be immune to the presence of noise in the PMU data or other noise elimination tools must be used in the protection scheme in order to make the protection scheme more efficient and reliable.

9.6   Summary

In this chapter, a broad overview of PMU application on the microgrid protection is provided. The presence of various DERs and other smart meters increases the complexity of the microgrid. Moreover, the microgrid environment is very dynamic nature. Therefore, the protection of the microgrid has been the primary concerns of the researchers in the last two decades. The dynamic nature of microgrid certainly demands synchronized data for the development of reliable protection scheme. In this aspect, the importance of the PMUs for microgrid protection is discussed in detail. The basic functionality, operation as well as requirement of PMU are discussed in different sections. An overview of microgrid protection with respect to various faults, islanding, as well as restoration scheme, is also provided in this chapter. The recent developments in microgrid protection using synchrophasor data are also provided with a proper literature survey. Finally, the future scope is also provided, where the issues such as time latency, missing data, and bad data are discussed. The researchers must investigate these common yet critical issues and study the impact of these issues on the developed protection scheme.

References

[1] A. G. Phadke and J. S. Thorp, "Synchronized phasor measurements and their applications," in *Power Electronics and Power Systems*, Springer, New York, NY, 2008.

[2] J. De La Ree, V. Centeno, J. S. Thorp, and A. G. Phadke, "Synchronized phasor measurement applications in power systems," *IEEE Transactions on Smart Grid*, vol. 1, no. 1, pp. 20–27, 2010.

[3] S. Nuthalapati and A. G. Phadke, "Managing the grid: using synchrophasor technology [guest editorial]," *IEEE Power & Energy Magazine*, vol. 13, no. 5, pp:10–12, 2015.

[4] M. Adibi and L. H. Fink, "Power system restoration planning," *IEEE Transactions on Power Systems*, vol. 9, no. 1, pp. 22–28, 1994.

[5] P. Kundu. and A. K. Pradhan., "Real-time analysis of power system protection schemes using synchronized data," *IEEE Transactions on Industrial Informatics*, vol.14, no. 9, pp. 3831–3839, 2018.

[6] A. M. Munoz, V. P. Lopez, J. J. González de la Rosa, *et al.*, "Embedding synchronized measurement technology for smart grid development," *IEEE Transactions on Industrial Informatics*, vol. 9, no. 1, pp. 52–61, 2013.

[7] Y. Liu, L. Zhan, Y. Zhang, *et al.*, "Wide-area measurement system development at the distribution level: an FNET/grid eye example," *IEEE Transactions on Power Delivery*, vol. 31, pp. 721–731, 2015.

[8] H. Mohsenian-Rad, E. Stewart, and E. Cortez, "Distribution synchrophasors: pairing big data with analytics to create actionable information," *IEEE Power & Energy Magazine*, vol. 16, no. 3, pp. 26–34, 2018.

[9] Y. Guo, K. Li, D. M. Laverty, and Y. Xue, "Synchrophasor based islanding detection for distributed generation systems using systematic principal component analysis approaches," *IEEE Transactions on Power Delivery*, vol. 30, no. 6, pp. 2544–2552, 2015.

[10] Z. Lin, T. Xia, Y. Ye, *et al.*, "Application of wide area measurement systems to islanding detection of bulk power systems," *IEEE Transactions on Power Systems*, vol. 28, no. 2, pp. 2006–2015, 2013.

[11] M. Almas and L. Vanfretti, "RT-HIL implementation of hybrid synchrophasor and GOOSE-based passive islanding schemes," *IEEE Transactions on Power Delivery*, vol. 31, pp. 1299–1309, 2015.

[12] R. M. R. A. Sankar and S. R. "Synchrophasor data driven islanding detection, localization and prediction for microgrid using energy operator," *IEEE Transactions on Power Systems*, vol. 36, no. 5, pp. 4052–4065, 2021.

[13] S. A. R. Konakalla, A. Valibeygi, and R. A. de Callafon, "Microgrid dynamic modeling and islanding control with synchrophasor data," *IEEE Transactions on Smart Grid*, vol. 11, no. 1, pp. 905–915, 2020.

[14] G. P. Kumar and P. Jena, "Pearson's correlation coefficient for islanding detection using Micro-PMU measurements," *IEEE Systems Journal*, vol. 15, no. 4, pp. 5078–5089, 2021.

[15] F. Ding, C. D. Booth, A. J. Roscoe, and S. Member, "Peak-ratio analysis method for enhancement of LOM protection using M-class PMUs," *IEEE Transactions on Smart Grid*, vol. 7, no. 1, pp. 291–299, 2016.

[16] M. A. Ebrahim, F. Wadie, and M. A. Abd-Allah, "An algorithm for detection of fault, islanding, and power swings in DG-equipped radial distribution networks," in *IEEE Systems Journal*, vol. 14, no. 3, pp. 3893–3903, 2020.

[17] D. M. Laverty, R. J. Best, and D. J. Morrow, "Loss-of-mains protection system by application of phasor measurement unit technology with experimentally assessed threshold settings," *IET Generation Transmission Distribution*, vol. 9, no. 2, pp. 146–153, 2015.

[18] A. Mishra and P. Jena, "A scheduled intentional islanding method based on ranking of possible islanding zone," *IEEE Transactions on Smart Grid*, vol. 12, no. 3, pp. 1853–1866, 2021, doi:10.1109/TSG.2020.3039384.

[19] Y. Lee, C. Chao, T. Lin, and C. Liu, "A synchrophasor-based fault location method for three-terminal hybrid transmission lines with one off-service line branch," *IEEE Transactions on Power Delivery*, vol. 33, no. 6, pp. 3249–3251, 2018.

[20] S. Azizi, M. Sanaye-Pasand, and M. Paolone, "Locating faults on untransposed, meshed transmission networks using a limited number of synchrophasor measurements," *IEEE Transactions on Power Systems*, vol. 31, no. 6, pp. 4462–4472, 2016, doi:10.1109/TPWRS.2016.2517185.

[21] N. K. Sharma and S. R. Samantaray, "PMU assisted integrated impedance angle-based microgrid protection scheme," *IEEE Transactions on Power Delivery*, vol. 35, no. 1, pp. 183–193, 2020, doi:10.1109/TPWRD.2019.2925887.

[22] Y. Seyedi, H. Karimi, S. Grijalva, J. Mahseredjian, and B. Sansò, "A supervised learning approach for centralized fault localization in smart microgrids," *IEEE Systems Journal*, vol. 16, pp. 4060–4070, 2021, doi:10.1109/JSYST.2021.3112710.

[23] N. K. Sharma and S. R. Samantaray, "A composite magnitude-phase plane of impedance difference for microgrid protection using synchrophasor measurements," *IEEE Systems Journal*, vol. 15, no. 3, pp. 4199–4209, 2021, doi:10.1109/JSYST.2020.2999483.

[24] P. K. Ganivada and P. Jena, "A fault location identification technique for active distribution system," *IEEE Transactions on Industrial Informatics*, vol. 18, no. 5, pp. 3000–3010, 2022.

[25] A. Gholami and F. Aminifar, "A hierarchical response-based approach to the load restoration problem," *IEEE Transactions on Smart Grid*, vol. 8, no. 4, pp. 1700–1709, 2017.

[26] Y. Zuo, G. Frigo, A. Derviškadić, and M. Paolone, "Impact of synchrophasor estimation algorithms in ROCOF-based under-frequency load-shedding," *IEEE Transactions on Power Systems*, vol. 35, no. 2, pp. 1305–1316, 2020.

[27] H. F. Habib, T. Youssef, M. H. Cintuglu, and O. A. Mohammed, "Multi-agent-based technique for fault location, isolation, and service restoration," *IEEE Transactions on Industry Applications*, vol. 53, no. 3, pp. 1841–1851, 2017.

[28] G. Frigo, A. Derviškadić, Y. Zuo, and M. Paolone, "PMU-based ROCOF measurements: uncertainty limits and metrological significance in power system applications," *IEEE Transactions on Instrumentation and Measurement*, vol. 68, no. 10, pp. 3810–3822, 2019.

[29] P. Demetriou, M. Asprou, and E. Kyriakides, "A real-time controlled islanding and restoration scheme based on estimated states," *IEEE Transactions on Power Systems*, vol. 34, no. 1, pp. 606–615, 2019.

[30] S. Yu, L. Zhang, H. H.-C. Lu, T. Fernando, and K. P. Wong, "A DSE-based power system frequency restoration strategy for PV-integrated power systems considering solar irradiance variations," *IEEE Transactions on Industrial Informatics*, vol. 13, no. 5, pp. 2511–2518, 2017.

[31] J. Xu, B. Xie, S. Liao, *et al.*, "Load shedding and restoration for intentional island with renewable distributed generation," *Journal of Modern Power Systems and Clean Energy*, vol. 9, no. 3, pp. 612–624, 2021.

[32] S. Konstantinopoulos, G. M. De Mijolla, J. H. Chow, H. Lev-Ari, and M. Wang, "Synchrophasor missing data recovery via data-driven filtering," *IEEE Transactions on Smart Grid*, vol. 11, no. 5, pp. 4321–4330, 2020.

[33] Z. Yang, H. Liu, T. Bi, Z. Li, and Q. Yang, "An adaptive PMU missing data recovery method," *International Journal of Electrical Power & Energy Systems*, vol. 116, Article no. 105577, 2020.

[34] P. Gao, M. Wang, S. G. Ghiocel, J. H. Chow, B. Fardanesh, and G. Stefopoulos, "Missing data recovery by exploiting low-dimensionality in power system synchrophasor measurements," *IEEE Transactions on Power Systems*, vol. 31, no. 2, pp. 1006–1013, 2016.

[35] W. Yuill, A. Edwards, S. Chowdhury, and S. P. Chowdhury, "Optimal PMU placement: a comprehensive literature review," in *2011 IEEE Power and Energy Society General Meeting*, 2011, pp. 1–8, doi:10.1109/PES.2011.6039376.

[36] F. J. Marin, F. Garcia-Lagos, G. Joya, and F. Sandoval, "Genetic algorithms for optimal placement of phasor measurement units in electric networks," *Electronics Letters*, vol. 39, no. 19, pp. 1403–1405, 2003.

[37] B. Milosevic and M. Begovic, "Nondominated sorting genetic algorithms for optimal phasor measurement placement," *IEEE Transactions on Power Systems*, vol. 18, no. 1, pp. 69–75, 2003.

[38] M. Hajain, A. M. Ranjbar, T. Amraee, and A. R. Shirani, "Optimal placement of phasor measurement units: particle swarm optimization approach," in *Proc. Int. Conf. Intelligent Systems and Applications in Power Systems.*, pp. 1–6, November 2007.

[39] J. Peng, Y. Sun, and H. F. Wang, "Optimal PMU placement for full network observability using Tabu search algorithm," *Electric Power Systems Research*, vol. 28, no. 4, pp. 223–231, 2006.

[40] D. Dua, S. Dambhare, R. K. Gajbhiye, and S. A. Soman, "Optimal multistage scheduling of PMU placement: an ILP approach," *IEEE Transactions on Power Delivery*, vol. 23, no. 4, pp. 1812–1820, 2008.

[41]   S. Chakrabarti, and E. Kyriakides, "Optimal placement of phasor measurement units for power system observability," *IEEE Transactions on Power Systems*, vol. 23, no. 3, pp. 1433–1440, 2008.

[42]   M. P. Anguswamy, M. Datta, L. Meegahapola, and A. Vahidnia, "Optimal micro-PMU placement in distribution networks considering usable zero-injection phase strings," *IEEE Transactions on Smart Grid*, vol. 13, no. 5, pp. 3662–3675, 2022, doi:10.1109/TSG.2022.3174917.

[43]   X. Chen, T. Chen, K. J. Tseng, Y. Sun, and G. Amaratunga, "Hybrid approach based on global search algorithm for optimal placement of μPMU in distribution networks," in *Proc. IEEE Innovat. Smart Grid Technol. Asia (ISGT-Asia)*, pp. 559–563, November/December, 2016.

[44]   M. A. Elgayar and N. H. Abbasy, "Optimal placement of micro PMUs in distribution networks using a graph theory/greedy hybrid algorithm," in *Proc. 21st Int. Middle East Power Syst. Conf. (MEPCON)*, pp. 1145–1149, December 2019.

[45]   R. Andreoni, D. Macii, M. Brunelli, and D. Petri, "Tri-objective optimal PMU placement including accurate state estimation: the case of distribution systems," *IEEE Access*, vol. 9, pp. 62102–62117, 2021, doi:10.1109/ACCESS.2021.3074579.

Index

Printed in the USA
CPSIA information can be obtained
at www.ICGtesting.com
JSHW062045171024
71825JS00002BA/8

9 781839 532849